General Relativity

General Relativity

A Concise Introduction

Steven Carlip
The University of California at Davis

OXFORD

Great Clarendon Street, Oxford, OX2 6DP,
United Kingdom

Oxford University Press is a department of the University of Oxford.
It furthers the University's objective of excellence in research, scholarship,
and education by publishing worldwide. Oxford is a registered trade mark of
Oxford University Press in the UK and in certain other countries

© Steven Carlip 2019

The moral rights of the author have been asserted

First Edition published in 2019

All rights reserved. No part of this publication may be reproduced, stored in
a retrieval system, or transmitted, in any form or by any means, without the
prior permission in writing of Oxford University Press, or as expressly permitted
by law, by licence or under terms agreed with the appropriate reprographics
rights organization. Enquiries concerning reproduction outside the scope of the
above should be sent to the Rights Department, Oxford University Press, at the
address above

You must not circulate this work in any other form
and you must impose this same condition on any acquirer

Published in the United States of America by Oxford University Press
198 Madison Avenue, New York, NY 10016, United States of America

British Library Cataloguing in Publication Data

Data available

Library of Congress Control Number: 2018957383

ISBN 978-0-19-882215-8 (hbk.)
ISBN 978-0-19-882216-5 (pbk.)

DOI: 10.1093/oso/9780198822158.001.0001

Links to third party websites are provided by Oxford in good faith and
for information only. Oxford disclaims any responsibility for the materials
contained in any third party website referenced in this work.

In memory of Bryce DeWitt and Cecile DeWitt-Morette

Preface

General relativity was born in 1915, the culmination of eight years of work, by Einstein and others, aimed at reconciling special relativity and Newton's action-at-a-distance gravity. In much of the century that followed, most physicists viewed the result with ambivalence. General relativity was seen as a beautiful, elegant theory, a model of what physics should be; but at the same time, it was a theory that seemed almost completely divorced from the rest of physics.

The beauty was obvious. Einstein had identified an assumption that had been taken for granted, that spacetime was flat and nondynamical; he had changed it in the simplest way possible; and out of that single step had sprung all of Newtonian gravity, small but measurable corrections, and a completely new view of cosmology. But the irrelevance also seemed obvious. General relativity remained stubbornly outside the quantum revolution that was sweeping through physics, and, as a practical matter, the best available technology had to be stretched to its limits to detect the tiny deviations from Newtonian gravity.

Things have changed. General relativity is still widely viewed as the model of elegance in physics, though some argue that its simplicity may be an accidental low energy manifestation of a more complicated high energy theory. But the separation from the rest of physics has ended. Cosmologists can no longer rely on a few simple solutions of the Einstein field equations; they must understand perturbations, gravitational lensing, and alternative theories of gravity. Gravitational waves are opening up an entirely new window into the Universe, allowing us to observe phenomena such as black hole mergers that would otherwise be completely invisible. High energy theorists find it increasingly difficult to escape the question of an "ultraviolet completion," a high energy limit that will almost certainly have to incorporate quantum gravity. Even some condensed matter and heavy ion physicists are looking at the peculiar links between their fields and gravity suggested by the AdS/CFT correspondence, and new research on evaporating black holes is pointing toward surprising connections between general relativity and quantum information theory.

This makes a difference in how we teach, and learn, general relativity. A course should be a jumping-off point for people going in many different directions. It shouldn't be mathematically sloppy—students will still need to read, and perhaps write, mathematically sophisticated papers—but it should move as quickly as possible to physics. It should tantalize, offering students glimpses of the vast landscape of science connected to general relativity without trying to explain everything at once.

This book has grown out of an introductory graduate course I've taught at the University of California at Davis since 1991. Davis operates on a quarter system—in practice, about 25 hours of instruction per course—and I've tried to write a textbook that could be used for such a course, although instructors with more time should find it easy to add material. The book is directed primarily toward graduate students, but my classes have often included a few undergraduates, who have kept up without too much extra work. I assume basic knowledge of special relativity (including four-vectors and Lorentz invariance), some familiarity with Lagrangians and variational principles, a reasonable level of comfort with partial differential equations and linear algebra, and an acquaintance with a small bit of set theory (open sets, intersections, and the like), but no prior knowledge of differential geometry or tensor analysis.

My assumption is that most students using this book will be physics students. I take the "physics first" approach, popularized by Jim Hartle, in which a class moves very quickly to calculations of gravity in the Solar System and only later returns to a more systematic development of the necessary mathematics. I have included a short introduction to the Hamiltonian formulation of general relativity, a topic often left out of introductory courses, and I close with a "bonus" chapter that briefly describes some of the many directions one could go from here.

I learned general relativity from Bryce DeWitt and Cecile DeWitt-Morette, to whom I owe a deep debt of gratitude. My students over the past 25 years taught me more. One of them, Joseph Mitchell, undertook a very careful reading of a draft. I thank my sons, Peter and David Carlip, for their help in editing and proofreading this book, and my brother, Walter Carlip, for extensive proofreading and invaluable help with typography. This work was supported in part by the U.S. Department of Energy under grant DE-FG02-91ER40674.

Contents

1	**Gravity as geometry**	**1**
	Further reading	3
2	**Geodesics**	**4**
	2.1 Straight lines and great circles	4
	2.2 Spacetime and causal structure	8
	2.3 The metric	9
	2.4 The geodesic equation	12
	2.5 The Newtonian limit	13
	Further reading	14
3	**Geodesics in the Solar System**	**15**
	3.1 The Schwarzschild metric	15
	3.2 Geodesics in the Schwarzschild metric	15
	3.3 Planetary orbits	18
	3.4 Light deflection	19
	3.5 Shapiro time delay	20
	3.6 Gravitational red shift	21
	3.7 The PPN formalism	23
	Further reading	24
4	**Manifolds and tensors**	**25**
	4.1 Manifolds	25
	4.2 Tangent vectors	27
	4.3 Cotangent vectors and gradients	29
	4.4 Tensors	31
	4.5 The metric	33
	4.6 Isometries and Killing vectors	34
	4.7 Orthonormal bases	35
	4.8 Differential forms	36
	Further reading	36
5	**Derivatives and curvature**	**37**
	5.1 Integration	37
	5.2 Why derivatives are more complicated	38
	5.3 Derivatives of p-forms: Cartan calculus	38
	5.4 Connections and covariant derivatives	39
	5.5 The Christoffel/Levi-Civita connection	42
	5.6 Parallel transport	44
	5.7 Curvature	45

x Contents

	5.8 Properties of the curvature tensor	47
	5.9 The Cartan structure equations	50
	Further reading	51

6 The Einstein field equations — 52
- 6.1 Gravity and geodesic deviation — 52
- 6.2 The Einstein-Hilbert action — 53
- 6.3 Conservation laws — 56
- 6.4 Generalizing the action — 57
- Further reading — 58

7 The stress-energy tensor — 59
- 7.1 Energy as a rank two tensor — 59
- 7.2 The stress-energy tensor for a point particle — 60
- 7.3 Perfect fluids — 61
- 7.4 Other fields — 61
- 7.5 Differential and integral conservation laws — 62
- 7.6 Conservation and the geodesic equation — 63
- Further reading — 65

8 The weak field approximation — 66
- 8.1 The linearized field equations — 66
- 8.2 The Newtonian limit — 68
- 8.3 Gravitomagnetism — 69
- 8.4 Higher orders — 69
- 8.5 Bootstrapping the field equations — 71
- Further reading — 71

9 Gravitational waves — 72
- 9.1 Solving the weak field equations — 72
- 9.2 Propagating waves — 74
- 9.3 Detecting gravitational waves — 76
- 9.4 Sources and signals — 77
- Further reading — 79

10 Black holes — 80
- 10.1 Static spacetimes — 80
- 10.2 Spherical symmetry — 81
- 10.3 Schwarzschild again — 82
- 10.4 The event horizon — 84
- 10.5 An extended spacetime — 86
- 10.6 Conformal structure and Penrose diagrams — 87
- 10.7 Properties of the horizon — 88
- Further reading — 91

11 Cosmology — 92
- 11.1 Homogeneity and isotropy — 92
- 11.2 The FLRW metric — 94

	11.3 Observational implications	96
	11.4 Inflation	98
	Further reading	100

12 The Hamiltonian formalism — 101
    12.1 The ADM metric — 101
    12.2 Extrinsic curvature — 102
    12.3 The Hamiltonian action — 104
    12.4 Constraints — 105
    12.5 Degrees of freedom — 105
    12.6 Boundary terms — 107
    Further reading — 108

13 Next steps — 109
    13.1 Solutions and approximations — 109
    13.2 Mathematical relativity — 110
    13.3 Alternative models — 112
    13.4 Quantum gravity — 112
    13.5 Experimental gravity — 115
    Further reading — 116

Appendix A   Mathematical details — 117
    A.1 Manifolds — 117
    A.2 Maps between manifolds — 118
    A.3 Topologies — 119
    A.4 Cohomology — 121
    A.5 Integrals of d-forms and Stokes' theorem — 122
    A.6 Curvature and holonomies — 123
    A.7 Symmetries and Killing vectors — 126
    A.8 The York decomposition — 127
    Further reading — 128

References — 129

Index — 135

1
Gravity as geometry

General relativity is an elegant and powerful theory, but it is also a strange one. According to Einstein, the phenomenon we usually think of as the force of gravity is really not a force at all, but rather a byproduct of the curvature of spacetime. Although we have become accustomed to this idea over time, it is still a peculiar notion, and it's worth trying to understand what it means before plunging into the details.

Start with the foundation of mechanics, Newton's second law,

$$\mathbf{F} = m\mathbf{a}. \tag{1.1}$$

Students who first see this equation sometimes worry that it might be a tautology. True, forces cause accelerations; but the way we recognize and measure a force is by the acceleration it causes. Is this not circular?

Looking deeper, though, we can see that the second law is rich with insight. Newton has split the Universe into two pieces: the right-hand side, the object whose motion we want to understand, and the left-hand side, the rest of the Universe, everything that might influence that motion. He tells us that the rest of the Universe causes accelerations, second derivatives of position—not velocities (first derivatives), not "jerks" (third derivatives), not anything else. This behavior is reflected throughout physics, where second order differential equations are found everywhere; it is only recently that we have begun to understand the underlying reason for this pattern (see Box 1.1).

Beyond this, Newton's second law tells us that the response of a body to the Universe has two separate elements: its acceleration, but also its inertia, or resistance to acceleration, as expressed by a single "inertial mass" m. This allows us to extract information about the force by measuring the response of otherwise identical bodies with different masses. So physics really is about forces.

But there is one exception. Consider the motion of an object in a gravitational field. Going back again to Newton—now to his law of gravity—we have

$$\not{m}\mathbf{a} = -\frac{GM\not{m}}{r^2}\hat{\mathbf{r}}. \tag{1.2}$$

The masses m on the left-hand and right-hand side cancel, and we are left only with acceleration.

This did not have to be. The mass m on the left-hand side of (1.2) is inertial mass, a measure of resistance to acceleration, while the mass m on the right-hand side is gravitational mass, a kind of "gravitational charge" analogous to electric charge. From the point of view of Newtonian gravity, there is no reason for these to be the

General Relativity: A Concise Introduction. Steven Carlip © Steven Carlip 2019.
Published in 2019 by Oxford University Press. DOI: 10.1093/oso/9780198822158.001.0001

Box 1.1 Ostrogradsky's theorem

The importance of acceleration in Newtonian mechanics is echoed throughout physics: almost all fundamental interactions are described by equations with at most two derivatives. We now understand that a good part of the explanation lies in an 1850 theorem of Ostrogradsky [1].

Starting with the Lagrangian formalism, Ostrogradsky explored the problem of constructing a Hamiltonian for a theory with more than two time derivatives. In an ordinary two-derivative theory, each generalized position q is accompanied by a generalized momentum p, which is related to the time derivative $\dot{q}$. For higher-derivative theories, new generalized momenta appear, related to higher derivatives of the generalized positions. Ostrogradsky showed that, with a clearly demarcated set of exceptions, the resulting Hamiltonians are linear in at least one of these new momenta. This means the energy is not bounded below: since the momentum can have any sign, there are configurations with arbitrarily negative energies. Such theories have no stable classical configurations, and no quantum ground states; they cannot describe a real world.

same. But the equality of these two masses—a form of what is called the equivalence principle—has been verified experimentally to better than a part in 10^{14}.

For gravity, then, physics is really not about forces, but about accelerations, or trajectories. Any sufficiently small "test body" will move along the same path in a gravitational field, no matter what its mass, shape, or internal composition. A gravitational field picks out a set of preferred paths.

But this is what we *mean* by a geometry. Euclidean geometry is ultimately a theory of straight lines; it consists of the statement, "These are the straight lines," and a description of their properties. Spherical geometry consists of the statement, "These are the great circles," and a description of their properties. If a gravitational field picks out a set of preferred "lines"—the paths of test bodies—and tells us their properties, it is determining a geometry.

The equivalence principle was already known to Galileo. The famous experiment in which he is said to have dropped two balls of different masses off the Leaning Tower of Pisa may not have taken place—it was first described years after his death—but he knew from many other experiments that bodies with different masses and compositions responded identically to gravity. Could he have formulated gravity as geometry?

Probably not: there is one more subtlety to take into account. In Euclidean geometry, two points determine a unique line. For an object moving in a gravitational field, on the other hand, the motion depends not only on the initial and final positions, but also on the initial velocity. I can drop a coin to the floor, or I can throw it straight up and allow it to fall. In either case, it will start and end at the same position.

But the coin will reach the floor at different *times*. If I specify the initial and final positions *and* times, the trajectory is unique. Gravity does, indeed, specify preferred

Box 1.2 Equivalence principles

The principle of equivalence has two common formulations. The first is "universality of free fall," the statement that the trajectory of a small object in a gravitational field is independent of its mass and internal properties such as structure and composition. The second is "weightlessness," the statement that in a small enough freely falling laboratory, no local measurement can detect the presence of the external gravitational field. The two formulations are essentially equivalent: we detected gravity by measuring accelerations, but universality of free fall implies the absence of relative accelerations in a freely falling laboratory.

There are nuances, though, depending on exactly what objects are included. The "weak equivalence principle" applies to objects whose internal gravitational fields can be neglected. The "Einstein equivalence principle" adds local position invariance and local Lorentz invariance, the statement that the outcome of any experiment in a freely falling reference frame is independent of the location and velocity of the frame. The "strong equivalence principle" extends these claims to objects whose self-gravitation cannot be ignored.

Precision tests of the weak equivalence principle date back to experiments by Eötvös in the early 20th century. Today, the principle has been tested to exquisite accuracy: relative accelerations of bodies with different compositions have been shown to be equal to about a part in 10^{14}. Progress has been made on testing the strong equivalence principle as well (see Box 8.1).

"lines," but the lines are not in space, they are in spacetime. Galileo and Newton didn't have special relativity—they had no unified treatment of space and time. Einstein did. Once that framework was available, the possibility of understanding gravity as geometry became available as well.

Further reading

A good introduction to the various forms of the principle of equivalence and their tests can be found in Will's book, *Theory and Experiment in Gravitational Physics* [2]. The first "modern" tests of the equivalence principle were performed by Eötvös in the early 1900s [3]. More up-to-date information on experimental tests can be found in, for instance, [4]. A beautiful review of Ostrogradsky's theorem is given in [5]. For a nonmathematical but deep and thought-provoking introduction to the idea of gravity as spacetime geometry, see Geroch's book, *General Relativity from A to B* [6].

2
Geodesics

Let us now look for a mathematical expression of this idea of gravity as geometry. To do so, we need to generalize the idea of a "straight line" to arbitrary geometries. In ordinary Euclidean geometry, a straight line can be characterized in either of two ways:

- A straight line is "straight," or autoparallel: its direction at any point is the same as its direction at any other. This turns out to be complicated to generalize—it requires us to know whether two vectors at different locations are parallel—and we will have to wait until Chapter 5 for a mathematical treatment.
- A straight line is the shortest distance between two points. This is much easier to generalize; it's exactly the kind of question that the calculus of variations was designed for.

This chapter will develop the general equation for the shortest, or more generally the extremal, line between two points, a "geodesic."

2.1 Straight lines and great circles

We start with the simplest example, a straight line in a two-dimension Euclidean space with Cartesian coordinates x and y. Choose initial and final points (x_0, y_0) and (x_1, y_1). Among all curves between these two points, we want to find the shortest.

To specify a curve, we *could* give y as a function of x, or x as a function of y. But this is contrary to the spirit of general relativity, in which all coordinates should be treated equally. For a more democratic description, we choose a parameter $\sigma \in [0, \sigma_{\max}]$, and describe the curve as a pair of functions $(x(\sigma), y(\sigma))$ with

$$(x(0), y(0)) = (x_0, y_0), \quad (x(\sigma_{\max}), y(\sigma_{\max})) = (x_1, y_1).$$

Mathematically, we are describing a curve as a map from the interval $[0, \sigma_{\max}]$ to $\mathbb{R}^2$. Physically, we can think of σ as a sort of "time" along the curve, though not the usual time coordinate, which we should treat as a coordinate like any other.

We next need the length of such a curve. If the curve is smooth enough, a short enough segment will be approximately straight, and we can use Pythagoras' theorem to write

$$ds^2 = dx^2 + dy^2 \tag{2.1}$$

for an "infinitesimal distance" ds, often referred to as a "line element." Mathematically inclined readers may be uncomfortable with this equation; we will see in Chapter 4 how to make it into a mathematically well-defined statement about covariant tensors.

General Relativity: A Concise Introduction. Steven Carlip © Steven Carlip 2019.
Published in 2019 by Oxford University Press. DOI: 10.1093/oso/9780198822158.001.0001

For now, we will take the physicists' approach and think of "ds" as representing a small but finite distance, taking limits when they make sense.

Let us define
$$E = \left(\frac{ds}{d\sigma}\right)^2 = \left(\frac{dx}{d\sigma}\right)^2 + \left(\frac{dy}{d\sigma}\right)^2. \tag{2.2}$$

The length of our curve is then
$$L = \int_{(x_0,y_0)}^{(x_1,y_1)} ds = \int_0^{\sigma_{\max}} E^{1/2} d\sigma. \tag{2.3}$$

We extremize by setting the variation to zero while holding the end points fixed:
$$\delta L = 0 = \int_0^{\sigma_{\max}} \frac{1}{2} E^{-1/2} \delta E \, d\sigma = \int_0^{\sigma_{\max}} E^{-1/2} \left(\frac{dx}{d\sigma}\frac{d\delta x}{d\sigma} + \frac{dy}{d\sigma}\frac{d\delta y}{d\sigma}\right) d\sigma. \tag{2.4}$$

It's helpful to remember that the variation δ is really just a derivative, albeit a derivative in an infinite-dimensional space of functions. In particular, the usual product rule and chain rule of calculus continue to hold. I've also used the fact that
$$\delta\left(\frac{dx}{d\sigma}\right) = \left(\frac{d\delta x}{d\sigma}\right);$$
when a variation changes $x(\sigma)$ to $x(\sigma) + \delta x(\sigma)$, the derivative changes accordingly.

To finish, we integrate (2.4) by parts to isolate δx and δy. In general, integration by parts leads to boundary terms, but here these terms are proportional to δx and δy, which are zero at the end points, since we are holding (x_0, y_0) and (x_1, y_1) fixed. Hence
$$0 = -\int_0^{\sigma_{\max}} \left[\frac{d}{d\sigma}\left(E^{-1/2}\frac{dx}{d\sigma}\right)\delta x + \frac{d}{d\sigma}\left(E^{-1/2}\frac{dy}{d\sigma}\right)\delta y\right] d\sigma. \tag{2.5}$$

Since the variations δx and δy are arbitrary, their coefficients must be zero, giving us the equations we seek.

One final trick simplifies this result. So far, σ has been an arbitrary label for points on the curve. Now that we have a particular curve, though, we can make the specific choice $\sigma = s$, labeling points by their distance from the initial point. (The reader may check that making this identification *before* the variation leads nowhere.) With this parametrization, $E = 1$, and the equations for the extremal curve are simply
$$\frac{d^2 x}{ds^2} = \frac{d^2 y}{ds^2} = 0, \tag{2.6}$$

the expected equations for a straight line.

This may seem to be a rather complicated way to get a simple result. The advantage is that it generalizes easily. Consider, for example, the same problem in polar coordinates,
$$x = r\cos\theta, \quad y = r\sin\theta. \tag{2.7}$$

Substituting $dx = \cos\theta \, dr - r\sin\theta \, d\theta$ and $dy = \sin\theta \, dr + r\cos\theta \, d\theta$ into the line element (2.1), we find

6 Geodesics

$$ds^2 = dr^2 + r^2 d\theta^2. \tag{2.8}$$

Now $E = \left(\dfrac{dr}{d\sigma}\right)^2 + r^2 \left(\dfrac{d\theta}{d\sigma}\right)^2$, and repeating the steps that led to (2.5), we have

$$0 = \int_0^{\sigma_{\max}} E^{-1/2} \left[\frac{dr}{d\sigma}\frac{d\delta r}{d\sigma} + r^2 \frac{d\theta}{d\sigma}\frac{d\delta\theta}{d\sigma} + r\left(\frac{d\theta}{d\sigma}\right)^2 \delta r\right] d\sigma$$

$$= -\int_0^{\sigma_{\max}} \left\{\left[\frac{d}{d\sigma}\left(E^{-1/2}\frac{dr}{d\sigma}\right) - E^{-1/2} r \left(\frac{d\theta}{d\sigma}\right)^2\right] \delta r + \frac{d}{d\sigma}\left[r^2 \frac{d\theta}{d\sigma}\right] \delta\theta\right\} d\sigma. \tag{2.9}$$

Note that E now depends explicitly on r, and not just its derivative; this factor of r must also be varied. Again choosing $\sigma = s$, we obtain

$$\frac{d}{ds}\left(r^2 \frac{d\theta}{ds}\right) = 0, \quad \frac{d^2 r}{ds^2} - r\left(\frac{d\theta}{ds}\right)^2 = 0. \tag{2.10}$$

These equations are easy enough to integrate, but we can also introduce a new trick. From (2.8), we have the identity

$$1 = \left(\frac{dr}{ds}\right)^2 + r^2 \left(\frac{d\theta}{ds}\right)^2. \tag{2.11}$$

This is not an independent equation—we will see below that it is a "first integral" of (2.10)—but it saves us a step.

Solving the system (2.10)–(2.11) is now straightforward. From (2.10),

$$r^2 \frac{d\theta}{ds} = b, \tag{2.12}$$

where b is an integration constant. Substituting into (2.11) and integrating, we obtain

$$r^2 = b^2 + (s - s_0)^2, \quad \theta - \theta_0 = \tan^{-1}\left(\frac{s - s_0}{b}\right), \tag{2.13}$$

and thus $r\cos(\theta - \theta_0) = b$. This is once again a straight line, as of course it must be.

Again, this may seem a roundabout way to find an obvious result. For a less trivial example, consider a "straight line" on a sphere of radius r_0, $x^2 + y^2 = r_0^2$. To obtain the line element, we start with flat three-dimensional space in spherical coordinates

$$x = r\sin\theta\cos\varphi, \quad y = r\sin\theta\sin\varphi, \quad z = r\cos\theta, \tag{2.14}$$

for which it is easy to check that

$$ds_3^2 = dx^2 + dy^2 + dz^2 = dr^2 + r^2 d\theta^2 + r^2 \sin^2\theta \, d\varphi^2. \tag{2.15}$$

We now restrict to the surface $r = r_0$, to obtain the line element for a sphere,

$$ds^2 = r_0^2 (d\theta^2 + \sin^2\theta \, d\varphi^2). \tag{2.16}$$

Box 2.1 Coordinates: part I

The line elements (2.1) and (2.8) look very different, but they describe the same flat space. Similarly, the geodesic equations (2.6) and (2.10) give different descriptions of the same straight lines. This is a first sign of a recurring theme in general relativity, the need to separate out the real physics from "coordinate effects."

For flat space, there is a preferred coordinate system, Cartesian coordinates. This can give us a false sense of security: we may think we understand what coordinates mean, especially if they're called x and y, or r and θ. But curved spacetimes typically have no preferred coordinates, and a good part of the work is to figure out what some chosen coordinates really mean. Much of the mathematics developed in Chapters 4 and 5 is designed to ensure that the final results don't depend on coordinates, but even then it can be tricky to interpret outcomes in terms of honest physical observables.

Proceeding as before, with $E = r_0^2 \left(\frac{d\theta}{d\sigma}\right)^2 + r_0^2 \sin^2\theta \left(\frac{d\varphi}{d\sigma}\right)^2$, we have

$$0 = \int_0^{\sigma_{\max}} E^{-1/2} r_0^2 \left[\frac{d\theta}{d\sigma}\frac{d\delta\theta}{d\sigma} + \sin^2\theta \frac{d\varphi}{d\sigma}\frac{d\delta\varphi}{d\sigma} + \sin\theta\cos\theta \left(\frac{d\varphi}{d\sigma}\right)^2 \delta\theta \right] d\sigma$$

$$= -r_0^2 \int_0^{\sigma_{\max}} \left\{ \left[\frac{d}{d\sigma}\left(E^{-1/2}\frac{d\theta}{d\sigma}\right) - E^{-1/2}\sin\theta\cos\theta \left(\frac{d\varphi}{d\sigma}\right)^2 \right]\delta\theta \right.$$
$$\left. + \frac{d}{d\sigma}\left[\sin^2\theta \frac{d\varphi}{d\sigma}\right] \delta\varphi \right\} d\sigma. \quad (2.17)$$

Setting $\sigma = s$, we obtain the equations

$$\frac{d}{ds}\left(\sin^2\theta \frac{d\varphi}{ds}\right) = 0, \quad \frac{d^2\theta}{ds^2} - \sin\theta\cos\theta \left(\frac{d\varphi}{ds}\right)^2 = 0, \quad (2.18)$$

and from (2.16), a first integral

$$1 = r_0^2 \left[\left(\frac{d\theta}{ds}\right)^2 + \sin^2\theta \left(\frac{d\varphi}{ds}\right)^2\right]. \quad (2.19)$$

The first equation in (2.18) gives

$$\sin^2\theta \frac{d\varphi}{ds} = b, \quad (2.20)$$

which allows us to integrate (2.19):

8 Geodesics

$$\cos\theta = (1 - r_0^2 b^2)^{1/2} \cos\left(\frac{s - s_0}{r_0}\right), \quad \tan(\varphi - \varphi_0) = -r_0 b \cot\left(\frac{s - s_0}{r_0}\right). \quad (2.21)$$

To interpret this solution, we can go back to the Cartesian coordinates (2.14). It is straightforward to check that

$$(\sin\varphi_0) x - (\cos\varphi_0) y - \left(\frac{r_0 b}{(1 - r_0^2 b^2)^{1/2}}\right) z = 0, \quad (2.22)$$

which may be recognized as the equation for a plane through the origin. The geodesics are thus intersections of the sphere with a plane through the origin. By definition, these are precisely the great circles.

There is a shortcut to this solution, which we will use in the next chapter. Note first that if we choose the integration constants $b = 1/r_0$, $\varphi_0 = \pi/2$, then (2.21) is the great circle around the equator, $\theta = \pi/2$, $\varphi = (s - s_0)/r_0$. This is a special case, but since the line element is spherically symmetric, we can always rotate our coordinates so that $\theta = \pi/2$ and $d\theta/ds = 0$ at the initial point $s = 0$. Then from (2.18)–(2.19),

$$\left.\frac{d^2\theta}{ds^2}\right|_{s=0} = \left.\frac{d^2\varphi}{ds^2}\right|_{s=0} = 0, \quad \left.\frac{d\theta}{ds}\right|_{s=0} = 0, \quad \left.\frac{d\varphi}{ds}\right|_{s=0} = \frac{1}{r_0}. \quad (2.23)$$

Since (2.18) is a second order system of differential equations, we are guaranteed that the higher derivatives add no information, so

$$\theta = \frac{\pi}{2}, \quad \varphi = \frac{s - s_0}{r_0} \quad (2.24)$$

is a solution for all s. By now undoing the rotation of coordinates, we can obtain any other great circle, reproducing the general solution.

2.2 Spacetime and causal structure

In the preceding section, we saw some simple examples of geodesics in flat and curved two-dimensional spaces. These examples were all *spatial* geodesics, though, geodesics in spaces whose line elements were positive definite. We are interested in spacetime, not just space, and this will require a further step.

To start, we need a generalization of the line element to spacetime. For the flat spacetime of special relativity, we know the answer. In special relativity, neither intervals in time alone nor in space alone have any invariant meaning. But as Einstein and Minkowski taught us, there is an invariant combination, the spacetime interval (or proper time)

$$ds^2 = dt^2 - \frac{1}{c^2}(dx^2 + dy^2 + dz^2), \quad (2.25)$$

also known as the "Minkowski metric." We will normally use units in which the speed of light is $c = 1$: if we measure time in seconds, we measure distance in light-seconds. We can thus write

$$ds^2 = dt^2 - dx^2 - dy^2 - dz^2. \quad (2.26)$$

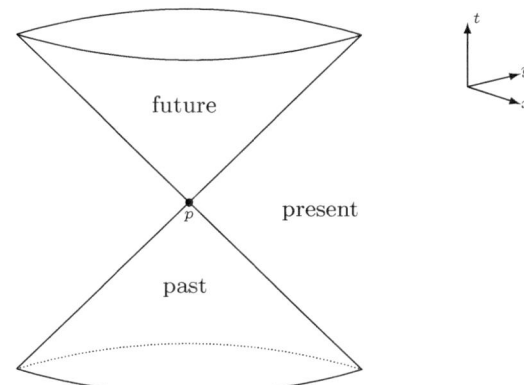

Fig. 2.1 The light cone of the point p. The expanding circular cross-sections of the future light cone, pictured in three dimensions, represent an expanding sphere of light in four dimensions.

The calculation of geodesics for the line element (2.26) is mathematically identical to the calculation in flat space, and the geodesics are again straight lines. The physics, though, is rather different. A geodesic can now have a "length" s that is positive real, zero, or imaginary. If ds^2 is positive, a geodesic, or more generally a segment of a curve, is "timelike." A simple example is the trajectory $x = y = z = 0$, $t = s$, for which an object only "moves" in time. If ds^2 is negative, the segment is "spacelike." A typical example is a curve in space at a fixed time $t = 0$. If $ds^2 = 0$, the segment is "lightlike" or "null." A typical example is the trajectory $x = ct$. More generally, a null curve is a possible path of a pulse of light, having "distance in space" equal to c times a "distance in time."

A path is causal if it is timelike or null everywhere. Such a path can be traversed by an object moving no faster than light. A point q is said to be in the future of the point p if there is future-directed causal path from p to q; it is in the past of p if there is a future-directed causal path from q to p. The light cone of p—the set of null curves passing through p—divides spacetime into three regions, the future, past, and present of p (see Fig. 2.1). The path of a massive object, called its "world line," always lies inside its light cone. This structure is familiar in special relativity, but it exists locally in curved spacetimes as well, although the global structure can be more complicated.

The terms spacelike, timelike, and null can also be applied to certain subspaces. Define a "hypersurface" to be a subspace of spacetime—technically, a submanifold—that has a dimension one less than the dimension of spacetime ($d = 3$ for our four-dimensional spacetime). A hypersurface is said to be spacelike if its normal vector is timelike, timelike if its normal is spacelike, and null if its normal is null.

2.3 The metric

We are now ready to tackle the problem of finding geodesics in an arbitrary curved spacetime. The starting point is again the line element. The lesson we take from

> **Box 2.2 Conventions: part I**
>
> A spacetime line element requires a choice of sign. The proper time (2.26) is positive for points with timelike separations (see Section 2.2) and negative for points with spacelike separations. But we could just as well have started with proper distance,
>
> $$ds^2 = -c^2 dt^2 + dx^2 + dy^2 + dz^2,$$
>
> which differs in sign. Option (2.26) is called the "mostly minuses" or "West Coast" convention; the opposite is "mostly pluses" or "East Coast." To describe motion in a gravitational field, the West Coast convention is convenient, since the parameter s is then proper time along a trajectory. Particle physicists also tend to prefer that convention, since it makes the square of the four-momentum positive on shell. In cosmology, on the other hand, the East Coast convention is easier, since the metric of space at constant time is then positive definite. With one possible obscure exception [7], either choice can be made. What matters, here and with other conventions, is consistency.

our examples—the hallmark of what mathematicians call Riemannian or pseudo-Riemannian geometry—is a version of Pythagoras' theorem, the principle that the square of the distance between two points is quadratic in the coordinate differences. In a d-dimensional spacetime, this means

$$ds^2 = \sum_{\mu,\nu=0}^{d-1} g_{\mu\nu} dx^\mu dx^\nu \qquad (2.27)$$

for some set of coordinates $\{x^\mu\}$ and some functions $g_{\mu\nu}(x)$. Note that the indices used to label coordinates in spacetime conventionally range from 0 to $d-1$, where x^0 is normally the "time" coordinate. The position of the indices—up for coordinates like x^μ, down for $g_{\mu\nu}$—encodes a mathematical distinction between tangent and cotangent vectors. We will explore this in Chapter 4; for now, it's just a notational convention. We can always assume that $g_{\mu\nu}$ is symmetric, that is, $g_{\mu\nu} = g_{\nu\mu}$, since it's easy to check that any additional antisymmetric piece would cancel out in the sum (2.27).

The functions $g_{\mu\nu}$ are called the components of the metric tensor in the coordinate system $\{x^\mu\}$. This rather long appellation is sometimes abbreviated to "the metric." The shortened terminology can be a bit misleading, though. We have seen that the line element in flat two-dimensional space may be written either as $dx^2 + dy^2$ in Cartesian coordinates or as $dr^2 + r^2 d\theta^2$ in polar coordinates. The components of the metric are clearly different, but the geometry is the same. We shall see in Chapter 4 that the metric tensor, properly defined, is independent of the choice of coordinates; we have merely expressed a single object in two different ways.

Equation (2.27), and many others, become simpler if we adopt the Einstein summation convention. This convention has two parts:

> **Box 2.3   Index gymnastics**
>
> A practitioner of general relativity must become adept at juggling indices. Here are a few tricks (make sure to work out where they come from)!
>
> - An index that is summed over is called a "dummy index." Any letter may be used for a dummy index, provided it isn't used elsewhere in the same term:
>
> $$A_\mu B^\mu = A_\nu B^\nu, \quad C_{\mu\nu}D^{\mu\nu} = C_{\nu\mu}D^{\nu\mu},$$
>
> since the sums consist of exactly the same terms.
> - Suppose $A_{\mu\nu}$ is antisymmetric in its indices, $S^{\mu\nu}$ is symmetric, and $T^{\mu\nu}$ has no particular symmetries. Then $A_{\mu\nu}S^{\mu\nu} = 0$ and
>
> $$A_{\mu\nu}T^{\mu\nu} = \frac{1}{2}A_{\mu\nu}(T^{\mu\nu} - T^{\nu\mu}), \quad S_{\mu\nu}T^{\mu\nu} = \frac{1}{2}S_{\mu\nu}(T^{\mu\nu} + T^{\nu\mu}).$$
>
> - If δ^μ_ν is the Kronecker delta, then $\delta^\mu_\nu A^\nu = A^\mu$ and $\delta^\mu_\mu = d$, where d is the dimension of the space or spacetime.

1. A given index can appear at most twice in any term of an equation. If it appears twice, this must be once as a lower index and once as an upper index.
2. If an index appears twice in a term, summation over that index is implied, and no explicit summation sign is needed.

Thus (2.27) becomes

$$ds^2 = g_{\mu\nu}dx^\mu dx^\nu. \tag{2.28}$$

The components $g_{\mu\nu}$ can be viewed as a symmetric $d \times d$ matrix, with the first index labeling the row and the second the column. Such a matrix has d eigenvalues. We assume that $g_{\mu\nu}$ is nondegenerate, that is, that the eigenvalues are all nonzero. The numbers p of positive eigenvalues and n of negative eigenvalues determine the "signature" of the metric, which can be denoted (p, n) or $p + n$, or, most commonly in physics, $(+ \ldots - \ldots)$. The line element (2.26), for instance, has signature $(1, 3)$ or $1 + 3$ or $(+ - - -)$. While it is not obvious, the signature is independent of the choice of coordinates. A line element with signature $(d, 0)$, like (2.8) or (2.16), is called Riemannian; one with signature $(1, d-1)$ or $(d-1, 1)$, like (2.26), is called Lorentzian.

The matrix inverse of $g_{\mu\nu}$, loosely referred to as the inverse metric, is written with upper indices, as $g^{\rho\sigma}$. Its existence is guaranteed by the nondegeneracy of the matrix of components. By definition, the inverse metric satisfies

$$g_{\mu\rho}g^{\rho\nu} = \delta^\nu_\mu, \tag{2.29}$$

where δ^ν_μ is the usual Kronecker delta. (Readers should check that this is, in fact, usual matrix multiplication.)

2.4 The geodesic equation

Starting with the general line element (2.28), we can now derive the geodesic equation for an arbitrary spacetime. The procedure is exactly the same as in Section 2.1:

$$E = g_{\mu\nu}\frac{dx^\mu}{d\sigma}\frac{dx^\nu}{d\sigma}, \qquad (2.30)$$

$$\delta L = 0 = \int_0^{\sigma_{max}} \frac{1}{2} E^{-1/2} \delta E \, d\sigma$$

$$= \int_0^{\sigma_{max}} \frac{1}{2} E^{-1/2} \left(g_{\mu\nu}\frac{d\delta x^\mu}{d\sigma}\frac{dx^\nu}{d\sigma} + g_{\mu\nu}\frac{dx^\mu}{d\sigma}\frac{d\delta x^\nu}{d\sigma} + \partial_\rho g_{\mu\nu}\frac{dx^\mu}{d\sigma}\frac{dx^\nu}{d\sigma}\delta x^\rho \right) d\sigma, \qquad (2.31)$$

where I have used the notation

$$\partial_\rho = \frac{\partial}{\partial x^\rho}. \qquad (2.32)$$

Note that the variation here is *not* a variation of the metric components $g_{\mu\nu}$, but rather of the path $x^\mu(\sigma)$. But "$g_{\mu\nu}$" in (2.30) means the value $g_{\mu\nu}(x(\sigma))$ on the path, so by the chain rule, when the path changes, $\delta g_{\mu\nu} = \delta x^\rho \partial_\rho g_{\mu\nu}$.

The first two terms in the second line of (2.31) are actually equivalent: they differ only by a renaming of the dummy indices μ to ν and ν to μ. Combining them, integrating by parts, and renaming some dummy indices, we see that

$$0 = -\int_0^{\sigma_{max}} \left[\frac{d}{d\sigma}\left(E^{-1/2} g_{\mu\nu}\frac{dx^\nu}{d\sigma}\right)\delta x^\mu - \frac{1}{2} E^{-1/2}\partial_\rho g_{\mu\nu}\frac{dx^\mu}{d\sigma}\frac{dx^\nu}{d\sigma}\delta x^\rho \right] d\sigma$$

$$= -\int_0^{\sigma_{max}} \left[\frac{d}{d\sigma}\left(E^{-1/2} g_{\rho\nu}\frac{dx^\nu}{d\sigma}\right) - \frac{1}{2} E^{-1/2}\partial_\rho g_{\mu\nu}\frac{dx^\mu}{d\sigma}\frac{dx^\nu}{d\sigma} \right] \delta x^\rho \, d\sigma. \qquad (2.33)$$

As before, the variation δx^ρ is arbitrary, so its coefficient must vanish:

$$\frac{d}{d\sigma}\left(E^{-1/2} g_{\rho\nu}\frac{dx^\nu}{d\sigma}\right) - \frac{1}{2} E^{-1/2}\partial_\rho g_{\mu\nu}\frac{dx^\mu}{d\sigma}\frac{dx^\nu}{d\sigma} = 0. \qquad (2.34)$$

Equation (2.34) is the general form of the geodesic equation. For spacelike and timelike geodesics, we can perform the same trick we used in Section 2.1, and choose $\sigma = s$. Then

$$\frac{d}{ds}\left(g_{\rho\nu}\frac{dx^\nu}{ds}\right) - \frac{1}{2}\partial_\rho g_{\mu\nu}\frac{dx^\mu}{ds}\frac{dx^\nu}{ds} = 0 \quad \text{(spacelike or timelike)}. \qquad (2.35)$$

For null geodesics, though, $s = 0$ and $E = 0$. Here we have two alternatives. We can return to the idea of a geodesic as an autoparallel line—we will do so in Chapter 5—or we can take a limit of (say) a timelike geodesic, defining a parameter

$$d\lambda = E^{1/2} d\sigma \qquad (2.36)$$

and letting E go to zero. In either case, the resulting equation is

$$\frac{d}{d\lambda}\left(g_{\rho\nu}\frac{dx^\nu}{d\lambda}\right) - \frac{1}{2}\partial_\rho g_{\mu\nu}\frac{dx^\mu}{d\lambda}\frac{dx^\nu}{d\lambda} = 0 \quad \text{(lightlike)}. \tag{2.37}$$

A parameter λ for which the geodesic equation takes this form is called an affine parameter.

There is another common form for the geodesic equation, obtained by differentiating the first term in (2.35) or (2.36) and using the product rule. The metric components $g_{\mu\nu}$ depend on s through their dependence on $x(s)$, so

$$\frac{d}{ds}g_{\mu\nu} = \partial_\rho g_{\mu\nu}\frac{dx^\rho}{ds}. \tag{2.38}$$

Using the inverse metric (2.29), it is easy to check that

$$\frac{d^2 x^\rho}{ds^2} + \Gamma^\rho_{\mu\nu}\frac{dx^\mu}{ds}\frac{dx^\nu}{ds} = 0 \tag{2.39}$$

with

$$\Gamma^\rho_{\mu\nu} = \frac{1}{2}g^{\rho\sigma}(\partial_\mu g_{\sigma\nu} + \partial_\nu g_{\sigma\mu} - \partial_\sigma g_{\mu\nu}). \tag{2.40}$$

The $\Gamma^\rho_{\mu\nu}$ are called the components of the Christoffel or the Levi-Civita connection, or sometimes the "Christoffel symbols"; in older papers they may be written as $\{{}^\rho_{\mu\nu}\}$.

Finally, let us consider the generalization of the first order equations (2.11) and (2.19). It is evident from the definition of the line element that

$$g_{\mu\nu}\frac{dx^\mu}{ds}\frac{dx^\nu}{ds} = 1 \quad \text{(spacelike or timelike)},$$

$$g_{\mu\nu}\frac{dx^\mu}{d\lambda}\frac{dx^\nu}{d\lambda} = 0 \quad \text{(lightlike)}. \tag{2.41}$$

These are not independent equations, though. Their first derivatives give back a linear combination of the geodesic equations: for instance,

$$\frac{d}{ds}\left(g_{\mu\nu}\frac{dx^\mu}{ds}\frac{dx^\nu}{ds}\right) = 2\frac{d}{ds}\left(g_{\mu\nu}\frac{dx^\nu}{ds}\right)\frac{dx^\mu}{ds} - \frac{dg_{\mu\nu}}{ds}\frac{dx^\mu}{ds}\frac{dx^\nu}{ds}$$

$$= 2\frac{dx^\rho}{ds}\left\{\frac{d}{ds}\left(g_{\rho\nu}\frac{dx^\nu}{ds}\right) - \frac{1}{2}\partial_\rho g_{\mu\nu}\frac{dx^\mu}{ds}\frac{dx^\nu}{ds}\right\}. \tag{2.42}$$

In that sense, (2.41) are "first integrals," obtained by integrating an appropriate combination of the geodesic equations, but with fixed integration constants determined from the definitions of the parameters s and λ.

2.5 The Newtonian limit

General relativity is a theory of physics, not mathematics, and we ought to check that this description of geodesics makes physical sense. A minimal test is that, at least to a very good approximation, we must be able to reproduce the remarkable successes of

Newtonian gravity. That is, we must be able to find a metric whose geodesics duplicate the trajectories that objects follow in a Newtonian gravitational field.

For this, we need the spacetime metric in the Newtonian approximation. Such a metric comes as a solution of the Einstein field equations, which we will not see until Chapter 6. For now, let us anticipate the result of Section 8.2. As one might expect, the approximate metric depends on the Newtonian gravitational potential Φ. It is, further, a "slow motion" solution—all speeds are much less than c—and a "weak field" solution—Φ/c^2 is of order v^2/c^2, as is typical for systems of gravitationally bound bodies. The line element then takes the approximate form

$$ds^2 = (1 + 2\Phi)dt^2 - (1 - 2\Phi)(dx^2 + dy^2 + dz^2). \tag{2.43}$$

The geodesic equations for this line element are simple. Equation (2.41) becomes

$$(1 + 2\Phi)\left(\frac{dt}{ds}\right)^2 = 1 + (1 - 2\Phi)\left[\left(\frac{dx}{ds}\right)^2 + \left(\frac{dy}{ds}\right)^2 + \left(\frac{dz}{ds}\right)^2\right], \tag{2.44}$$

while (2.35) becomes

$$-\delta_{ij}\frac{d}{ds}\left((1-2\Phi)\frac{dx^j}{ds}\right) - \partial_i\Phi\left[\left(\frac{dt}{ds}\right)^2 + \left(\frac{dx}{ds}\right)^2 + \left(\frac{dy}{ds}\right)^2 + \left(\frac{dz}{ds}\right)^2\right] = 0$$
$$(i = 1, 2, 3). \tag{2.45}$$

If we now neglect terms of order v^2/c^2—the order at which ordinary special relativistic corrections become important—these equations reduce to

$$\frac{dt}{ds} = 1, \quad \delta_{ij}\frac{d^2x^j}{dt^2} = -\partial_i\Phi \quad (i = 1, 2, 3), \tag{2.46}$$

which are just the Newtonian equations of motion for an object in a gravitational field. The appropriate spacetime geometry can thus reproduce Newtonian motion.

Further reading

The geodesic equation can also be derived by means of the usual Euler-Lagrange equations: see, for example, Section 7.6 of d'Inverno's *Introducing Einstein's Relativity* [8]. An introduction to the line element (2.26) in special relativity, focusing on its geometric significance, may be found in Taylor and Wheeler's classic, *Spacetime Physics* [9], or in Chapter 1 of the textbook by Schutz, *A First Course in General Relativity* [10]. The inside cover of Misner, Thorne, and Wheeler's famous textbook *Gravitation* [11] has a table of conventions used by various authors; while somewhat outdated, it gives an overview of the variety of choices.

3
Geodesics in the Solar System

At the end of the preceding chapter, we saw that general relativity can successfully reproduce the equations of motion of Newtonian gravity. It does more than that, of course: even in the Solar System, it predicts small but observable corrections to Newtonian physics. The four "classical tests" of general relativity—the precession of planetary orbits, the deflection of light by a gravitational field, the gravitational time delay of light, and gravitational red shift and time dilation—can also be obtained from the geodesic equation.

3.1 The Schwarzschild metric

To investigate these predictions, we need a line element that is more accurate than the weak field approximation (2.43). The spacetime geometry outside a nonrotating spherical mass was first worked out by Karl Schwarzschild in 1916, very soon after Einstein published the field equations of general relativity. We will derive the resulting Schwarzschild metric in Chapter 10, where we will obtain a line element

$$ds^2 = \left(1 - \frac{2m}{r}\right) dt^2 - \left(1 - \frac{2m}{r}\right)^{-1} dr^2 - r^2(d\theta^2 + \sin^2\theta d\varphi^2). \tag{3.1}$$

Note that when $m = 0$, this reduces to the flat Minkowski metric (2.26) in the spherical coordinates of eqn (2.15). Similarly, as r approaches infinity, the metric approaches the Minkowski metric; the Schwarzschild metric is "asymptotically flat."

The parameter m is proportional to the mass of the gravitating object,

$$m = \frac{GM}{c^2}, \tag{3.2}$$

where M is the ordinary mass and G is Newton's gravitational constant. The quantity m has units of length, with a value of about 4.5 mm for the Earth and 1.5 km for the Sun. The ratio m/r in (3.1) is therefore typically very small; even at the surface of the Sun, it is only about 2×10^{-6}. The Schwarzschild line element for the Solar System thus deviates only very slightly from the line element of flat spacetime. Nevertheless, this small difference is responsible for the motion of the planets.

3.2 Geodesics in the Schwarzschild metric

Let us now calculate the geodesics for the line element (3.1). We will start with the timelike geodesics that describe the motion of planets, and then turn to the null

General Relativity: A Concise Introduction. Steven Carlip © Steven Carlip 2019.
Published in 2019 by Oxford University Press. DOI: 10.1093/oso/9780198822158.001.0001

Box 3.1 Coordinates: part II

It is tempting to think of the coordinate r as the familiar radial coordinate of flat space. For large r, this is a good approximation, since the line element is nearly flat. For small r, though, this picture breaks down. As we shall see in Chapter 10, the flat space coordinate r serves many different functions—it is, for example, both a radial distance and a square root of the area of a sphere. In a curved spacetime, these are no longer equivalent, and different choices lead to different forms of the line element.

For instance, in (3.1), coordinate distances in the radial direction differ from those in angular directions. But if we transform to "isotropic coordinates,"

$$r = \bar{r}\left(1 + \frac{m}{2\bar{r}}\right)^2, \tag{3.4}$$

the line element becomes

$$ds^2 = \frac{\left(1 - \frac{m}{2\bar{r}}\right)^2}{\left(1 + \frac{m}{2\bar{r}}\right)^2}dt^2 - \left(1 + \frac{m}{2\bar{r}}\right)^4 (d\bar{r}^2 + \bar{r}^2(d\theta^2 + \sin^2\theta d\varphi^2))$$

$$= \frac{\left(1 - \frac{m}{2\bar{r}}\right)^2}{\left(1 + \frac{m}{2\bar{r}}\right)^2}dt^2 - \left(1 + \frac{m}{2\bar{r}}\right)^4 (dx^2 + dy^2 + dz^2), \tag{3.5}$$

with "Cartesian" coordinates x, y, and z that contribute symmetrically to isotropic distance $\bar{r} = (x^2 + y^2 + z^2)^{1/2}$. At large distances, the difference between r and $\bar{r}$ is insignificant, and to first order in m/r the predictions for phenomena such as the deflection of light are identical. At the next order, though, the equations depend on which "r" we use. This is not a true physical difference, but to see this takes some careful work [12]: we have to reexpress the results in terms of unambiguously measurable coordinate-independent quantities.

geodesics that describe the paths of light rays. From (3.1), the nonzero components of the metric are

$$g_{tt} = 1 - \frac{2m}{r}, \quad g_{rr} = -\left(1 - \frac{2m}{r}\right)^{-1}, \quad g_{\theta\theta} = -r^2, \quad g_{\varphi\varphi} = -r^2 \sin^2\theta, \tag{3.3}$$

where the index labels t, r, θ, φ mean the same thing as $0, 1, 2, 3$. In writing out the geodesic equation, it's helpful to keep two shortcuts in mind:

- The metric is diagonal, so in sums like $g_{\rho\mu}\frac{dx^\mu}{ds}$ only terms with $\mu = \rho$ appear.
- The components (3.3) depend only on r and θ, so $\partial_\mu g_{\sigma\tau}$ is zero unless μ is r or θ.

We now look at the four equations (2.35) and the first integral (2.41):

Geodesics in the Schwarzschild metric

$\mu = t:$ $\quad \dfrac{d}{ds}\left(g_{t\mu}\dfrac{dx^\mu}{ds}\right) - \dfrac{1}{2}\partial_t g_{\mu\nu}\dfrac{dx^\mu}{ds}\dfrac{dx^\nu}{ds} = 0 = \dfrac{d}{ds}\left[\left(1-\dfrac{2m}{r}\right)\dfrac{dt}{ds}\right],$ (3.6)

$\mu = r:$ complicated expression; we'll use the first integral instead,

$\mu = \theta:$ $\quad \dfrac{d}{ds}\left(g_{\theta\mu}\dfrac{dx^\mu}{ds}\right) - \dfrac{1}{2}\partial_\theta g_{\mu\nu}\dfrac{dx^\mu}{ds}\dfrac{dx^\nu}{ds} = 0$

$$= -\dfrac{d}{ds}\left(r^2 \dfrac{d\theta}{ds}\right) + r^2 \sin\theta \cos\theta \left(\dfrac{d\varphi}{ds}\right)^2, \quad (3.7)$$

$\mu = \varphi:$ $\quad \dfrac{d}{ds}\left(g_{\varphi\mu}\dfrac{dx^\mu}{ds}\right) - \dfrac{1}{2}\partial_\varphi g_{\mu\nu}\dfrac{dx^\mu}{ds}\dfrac{dx^\nu}{ds} = 0 = -\dfrac{d}{ds}\left(r^2 \sin^2\theta \dfrac{d\varphi}{ds}\right),$ (3.8)

first integral: $\quad 1 = \left(1-\dfrac{2m}{r}\right)\left(\dfrac{dt}{ds}\right)^2 - \left(1-\dfrac{2m}{r}\right)^{-1}\left(\dfrac{dr}{ds}\right)^2$

$$- r^2 \left(\dfrac{d\theta}{ds}\right)^2 - r^2 \sin^2\theta \left(\dfrac{d\varphi}{ds}\right)^2. \quad (3.9)$$

Let us start with (3.7). As in Section 2.1, we can rotate our coordinates so that at an initial time, $\theta = \pi/2$ and $d\theta/ds = 0$. Then (3.7) implies that

$$\theta = \dfrac{\pi}{2} \quad (3.10)$$

for all times. Just as in Newtonian gravity, trajectories remain in a single plane.

Next, we can integrate (3.6) and 3.8):

$$\left(1-\dfrac{2m}{r}\right)\dfrac{dt}{ds} = \tilde{E}, \quad (3.11)$$

$$r^2 \dfrac{d\varphi}{ds} = \tilde{L}, \quad (3.12)$$

where $\tilde{E}$ and $\tilde{L}$ are integration constants. (The tildes are a convention to distinguish timelike geodesics, the case here, from null geodesics, which will come later.) We are left with (3.9), which now simplifies to

$$\left(\dfrac{dr}{ds}\right)^2 = \tilde{E}^2 - \left(1-\dfrac{2m}{r}\right)\left(1+\dfrac{\tilde{L}^2}{r^2}\right). \quad (3.13)$$

Equation (3.13) is integrable, and in principle (3.10)–(3.13) give a complete description of the geodesics of the Schwarzschild geometry. But the results can only be expressed in terms of certain special functions, elliptic integrals, that are unfamiliar to most physicists. So even though exact solutions are available, it is useful to find approximations that involve more easily understood functions.

3.3 Planetary orbits

To find these approximate solutions, we start with two tricks from Newtonian celestial mechanics. We write the equations in terms of a new variable $u = 1/r$, and we replace derivatives with respect to s with derivatives with respect to φ,

$$\frac{du}{ds} = \frac{du}{d\varphi}\frac{d\varphi}{ds} = \tilde{L}u^2 \frac{du}{d\varphi}, \tag{3.14}$$

where the second equality follows from (3.12). Equation (3.13) then becomes

$$\left(\frac{du}{d\varphi}\right)^2 = \frac{\tilde{E}^2 - 1}{\tilde{L}}^2 + \frac{2m}{\tilde{L}^2}u - u^2 + 2mu^3. \tag{3.15}$$

Differentiating with respect to φ, we obtain either $\dfrac{du}{d\varphi} = 0$—the condition for a circular orbit—or

$$\frac{d^2u}{d\varphi^2} + u - 3mu^2 = \frac{m}{\tilde{L}^2}. \tag{3.16}$$

Now, recall that for motion in the Solar System, $mu = m/r \ll 1$. As a first approximation, we can therefore neglect the term $3mu^2$ in (3.16). The equation is then trivial to solve:

$$u = \frac{m}{\tilde{L}^2} + \frac{1}{d}\cos(\varphi - \varphi_0), \tag{3.17}$$

where d is an integration constant. This may be recognized as the equation for a conic section—an ellipse, parabola, or hyperbola—with directrix d and eccentricity $e = \tilde{L}^2/md$. For an ellipse, in particular, the semimajor axis is $a = de/(1 - e^2)$. (It's a nice exercise to convert (3.17) to Cartesian coordinates to check that it represents a conic section.) We thus reproduce the Newtonian result for motion in the gravitational field of a central mass.

There are several ways to refine this approximation. One simple method is to assume a nearly circular orbit—a good approximation for all of the planets in the Solar System—and write $u = u_0 + v$ with u_0 constant and v small. Then (3.16) becomes

$$\frac{d^2v}{d\varphi^2} + (1 - 6mu_0)v - 3mv^2 = \frac{m}{\tilde{L}^2} - u_0 + 3mu_0^2. \tag{3.18}$$

We can choose u_0 to make the right-hand side vanish. If we neglect the term $3mv^2$, which is much smaller than the term $3mu^2$ we dropped previously, we obtain

$$u = u_0 + \frac{1}{d}\cos\left(\sqrt{1 - 6mu_0}\,(\varphi - \varphi_0)\right). \tag{3.19}$$

For a bound orbit, this is nearly an ellipse, but the curve now doesn't quite close. If we choose an initial value for u—for instance, the perihelion, or closest approach to the Sun—the orbit will return to that value after an angular change of

$$\Delta\varphi = \frac{2\pi}{\sqrt{1 - 6mu_0}} \approx 2\pi + \frac{6\pi m}{r_0}. \tag{3.20}$$

This is the famous precession of the perihelion, the first crucial test of general relativity. Astronomers had known since 1859 that Mercury's orbit did not quite match the predictions of Newtonian gravity, and various ad hoc solutions had been proposed, from an unknown inner planet ("Vulcan") to a modification of the inverse square law ($r^{-2.0000001574}$ would do the job). The correction from general relativity settled the problem, exactly matching the anomalous precession, currently to an accuracy of about 10^{-4}.

Within the Solar System, the anomalous precession is largest for Mercury, which has the smallest r_0, but the predictions have also been tested for Venus, Earth, and Mars; for the asteroid Icarus; for the Earth-orbiting satellite LAGEOS II; for a number of binary stellar systems; and for binary pulsar systems, in which the precession can be as large as several degrees per year.

3.4 Light deflection

The next two classical tests of general relativity involve light. For this, we need to solve the equation for null geodesics, (2.37) rather than (2.35). Fortunately, the calculations of Section 3.2 are almost unaltered: we just have to change our parameter from s to λ and to replace the 1 in the left-hand side of (3.9) with 0. To avoid confusion, we will also replace the integration constants $\tilde{E}$ and $\tilde{L}$ with E and L.

There is an important physical difference, though. For planetary orbits, we observe the distance from the Sun as a function of angle or time. For light, on the other hand, we observe the angle at which a ray approaches us. This means that for planetary orbits, we need u as a function of φ, while for light we need φ as a function of u. Mathematically, of course, these are equivalent, but the physics dictates which form is more useful.

It is now easy to check that for null geodesics, (3.15) becomes

$$\left(\frac{d\varphi}{du}\right)^2 = \frac{1}{\frac{1}{b^2} - u^2(1-2mu)} \qquad (3.21)$$

with $b^2 = \frac{L^2}{E^2}$, that is,

$$\varphi = \pm \int \frac{du}{\left[\frac{1}{b^2} - u^2(1-2mu)\right]^{1/2}}. \qquad (3.22)$$

The quantity b is approximately the impact parameter, the distance of closest approach of the light ray to the gravitating body.

The sign in (3.22) matters. As a ray of light approaches the Sun from $\varphi = 0$, the angle increases while $u = \frac{1}{r}$ increases. After its closest approach at $u \approx \frac{1}{b}$, the angle φ continues to increase, but u decreases, so the derivative switches sign. In principle, we should break the integral (3.22) into two pieces with opposite signs and join the two solutions. In practice, though, if we are looking at light propagating from far away ($u \sim 0$) to far away ($u \sim 0$), the set-up is symmetric, so it is enough to calculate the deflection on one leg and then double it.

The integral (3.22) is again an unfamiliar elliptic integral, so let us once more look for a more transparent approximation. As a first step, we again neglect the term $2mu$ in the denominator, yielding

$$u = \frac{1}{b}\sin(\varphi - \varphi_0) \Leftrightarrow r\sin(\varphi - \varphi_0) = b, \qquad (3.23)$$

the equation for a straight line. As a next approximation, let us set

$$v = u(1 - mu), \quad dv = (1 - 2mu)\, du = (1 - 4mv)^{1/2}\, du. \qquad (3.24)$$

If we now neglect terms of order $(mv)^2$—which is much smaller than mu—the positive branch of the integral (3.22) becomes

$$\varphi = \int \frac{(1+2mv)dv}{\left(\frac{1}{b^2} - v^2\right)^{1/2}} \Rightarrow \varphi = \sin^{-1}(bv) - 2m\left(\frac{1}{b^2} - v^2\right)^{1/2} + \frac{2m}{b}, \qquad (3.25)$$

where the constant of integration is chosen so that $\varphi = 0$ when $v = 0$. If a ray of light starts infinitely far away with $\varphi = 0$, the angle will increase to a maximum value of $\varphi = \frac{\pi}{2} + \frac{2m}{b}$ at the point of nearest approach, $v = \frac{1}{b}$. By symmetry, the deflection will be the same on the negative branch of the integral as the ray moves away. The total change in the angle will thus be

$$\Delta\varphi = \pi + \frac{4m}{b}, \qquad (3.26)$$

a deflection of $\frac{4m}{b}$ from a straight line. Note that this deflection is achromatic—the angle is independent of wavelength, making it very different from other possible sources of light-bending.

To measure this deflection with visible light, one must observe stars near the limb of the Sun (small b), which means waiting for a Solar eclipse to allow the stars to be seen. The first observations, by a 1919 expedition led by the British astronomer Arthur Eddington, confirmed the predictions (3.26) to an accuracy of about 30%. (The *New York Times* front page headline read, "Lights all askew in the heavens. Men of science more or less agog over results of eclipse observations.") Subsequent eclipse observations increased the accuracy, but with persistent problems; the Texas-Mauritania expedition of 1973, for instance, was hit by a sandstorm.

With the advent of radio interferometry, though, and especially the use of very precise Very-Long-Baseline Interferometry (VLBI), we can measure the deflection of radio waves from quasars without waiting for an eclipse. As a result, the deflection (3.26) is now confirmed to an accuracy of about 10^{-4}. Deflection of light by Jupiter has also been observed. Beyond the Solar System, bending of light by galaxies and clusters of galaxies—leading to multiple images of stars, "Einstein rings," and more subtle shape distortions by weak gravitational lensing—has become an important tool for cosmology.

3.5 Shapiro time delay

A massive body such as the Sun not only deflects light, but also delays it. This effect, the Shapiro time delay, was only identified in 1964, although in retrospect it could

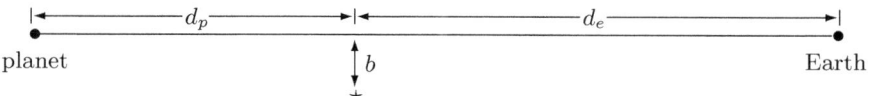

Fig. 3.1 Shapiro time delay for a ray of light passing the Sun

have been found much earlier. The time delay is distinct from deflection; although the deflection of light slightly lengthens its path, that change is negligible. The much larger Shapiro delay is, rather, a direct consequence of the effect of spacetime curvature on distance and time.

The Shapiro time delay can be calculated from the Schwarzschild line element (3.1), but it is simpler to understand in Cartesian-like coordinates, and the Newtonian approximation (2.43) is good enough. In this approximation, a ray of light will travel in a straight line, say in the x direction. Such a trajectory has $y = y_0$, and $z = z_0$, and the impact parameter—the distance of closest approach to the origin—is

$$b = (x_0^2 + y_0^2)^{1/2}.$$

The Newtonian potential for a spherical mass at the origin is $\Phi = -\frac{m}{r}$, so from (2.43), a null geodesic satisfies

$$ds^2 = 0 = \left(1 - \frac{2m}{r}\right) dt^2 - \left(1 + \frac{2m}{r}\right) dx^2. \tag{3.27}$$

Thus, to first order in the small quantity m/r,

$$t = \int \left(1 + \frac{2m}{r}\right) dx = x + 2m \sinh^{-1}\left(\frac{x}{b}\right) \approx x + 2m \ln \frac{2|x|}{b}. \tag{3.28}$$

In particular, for a ray of light passing the Sun, the elapsed time will be slightly longer than the Newtonian value of x/c. As in the preceding section, we have to be careful to distinguish the path along which the ray approaches the Sun and the path along which it moves away. But it is easy to check that for the configuration shown in Fig. 3.1—light traveling from a planet past the Sun to the Earth—the two legs add, and the time delay is

$$\Delta t = 2m \ln \frac{4 d_p d_e}{b^2}. \tag{3.29}$$

Irwin Shapiro discovered this effect in the early 1960s, in the first years of planetary radar ranging. The first tests, in 1966-67, used radar signals bounced off Mercury and Venus, and confirmed (3.29) at an accuracy of about 20%. Subsequent observations of planets and satellites have pushed the uncertainty down to a few times 10^{-5}.

3.6 Gravitational red shift

The final classical test of general relativity is the gravitational red shift. Consider two observers at rest in the Schwarzschild spacetime, at fixed spatial positions r_1 and r_2,

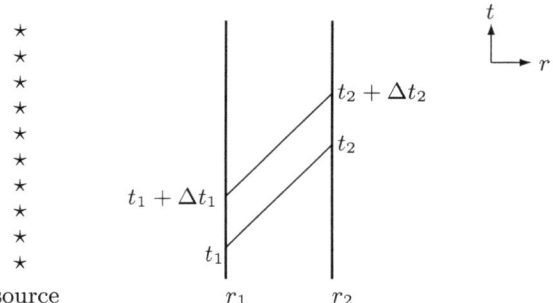

Fig. 3.2 Gravitational red shift

with $\theta_1 = \theta_2$ and $\varphi_1 = \varphi_2$. Suppose, as illustrated in Fig. 3.2, that the observer at r_1 sends a pulse of light towards r_2 at time t_1, and a second pulse at time $t_1 + \Delta t_1$. We can think of these pulses as two successive peaks of a single light wave; Δt_1 is then a measure of the period of the wave.

Again, light travels along null geodesics,

$$ds^2 = \left(1 - \frac{2m}{r}\right) dt^2 - \left(1 - \frac{2m}{r}\right)^{-1} dr^2 = 0 \Rightarrow dt = \left(1 - \frac{2m}{r}\right)^{-1} dr. \quad (3.30)$$

Hence for the two pulses of light,

$$t_2 - t_1 = \int_{r_1}^{r_2} \left(1 - \frac{2m}{r}\right)^{-1} dr,$$

$$(t_2 + \Delta t_2) - (t_1 + \Delta t_1) = \int_{r_1}^{r_2} \left(1 - \frac{2m}{r}\right)^{-1} dr, \quad (3.31)$$

and thus $\Delta t_2 = \Delta t_1$. The *coordinate* time between successive pulses, or successive peaks of a light wave, doesn't change.

But coordinates aren't physical—they are just arbitrary human-made labels. The time *measured* by an observer is the proper time s. In particular, for an observer at rest at position r in a Schwarzschild spacetime, it follows directly from the Schwarzschild line element that

$$\Delta s = \left(1 - \frac{2m}{r}\right)^{1/2} \Delta t. \quad (3.32)$$

In the set-up of Fig. 3.2, therefore,

$$\frac{\Delta s_2}{\Delta s_1} = \left(1 - \frac{2m}{r_2}\right)^{1/2} \left(1 - \frac{2m}{r_1}\right)^{-1/2} \approx 1 + \frac{m}{r_1} - \frac{m}{r_2}. \quad (3.33)$$

We can translate this into a statement about frequency: since the (angular) frequency ω is inversely proportional to the period,

$$\frac{\omega_1 - \omega_2}{\omega_2} \approx \frac{m}{r_1} - \frac{m}{r_2}. \quad (3.34)$$

In particular, if $r_2 > r_1$—if light is "climbing out of a gravitational well"—then $\omega_2 < \omega_1$. This is the gravitational red shift. More generally, the relation (3.33) illustrates gravitational time dilation: a clock in a stronger gravitational field will be seen as running slow by an observer in a weaker field.

Einstein actually predicted the gravitational red shift in 1907, before the final formulation of general relativity. Early attempts to see the red shift in the spectra of white dwarf stars were largely inconclusive. The first unambiguous observation was performed by Robert Pound and Glen Rebka in 1960. Using the Mossbauer effect to obtain gamma rays of a very precise frequency, they verified (3.34) to an accuracy of about 10%. Subsequent observations, most notably the Gravity Probe A launch of a hydrogen maser into space, have pushed the uncertainty down to about 2×10^{-4}, while planned experiments should gain another two orders of magnitude of accuracy.

Gravitational time dilation can now be measured directly by comparing atomic clocks at different heights, over distances as small as a third of a meter. This effect is also one of the few cases in which general relativity impinges upon daily life. The atomic clocks on Global Positioning System (GPS) satellites gain about 46 microseconds per day relative to Earth clocks because of gravitational time dilation, while losing 7 microseconds per day from special relativistic time dilation. If these effects were not accounted for, the system would very quickly become useless.

(In fact, when the NTS-2 satellite, the precursor of the GPS system, was launched in 1977, some engineers questioned the relevance of these relativistic effects. So the satellite was launched without corrections, but with a frequency synthesizer that could be switched on from the ground to correct the clock rates if needed. After 20 days, the relativistic effects were confirmed to an accuracy of about 1%, and the corrections were turned on.)

3.7 The PPN formalism

The four "classical tests" are predictions of general relativity that can be directly compared with observation. In each case, nature behaves as predicted. But it would be useful to do more, to combine these observations to see just how tightly the theory is constrained. For weak fields, the parametrized post-Newtonian (PPN) formalism provides a framework to do this.

To see how this works, let us start with the Schwarzschild spacetime in isotropic coordinates (3.5) and expand in powers of $m/\bar{r}$, but now with two new parameters β and γ:

$$ds^2 = \left[1 - \frac{2m}{\bar{r}} + 2\beta\left(\frac{m}{\bar{r}}\right)^2\right] dt^2 - \left(1 + 2\gamma\frac{m}{\bar{r}}\right)(dx^2 + dy^2 + dz^2). \tag{3.35}$$

One could imagine a third parameter in the coefficient of $m/\bar{r}$ in g_{tt}, but its value is fixed by the Newtonian limit of Section 2.5. For general relativity, $\beta = \gamma = 1$; as we shall see in Chapter 10, these values are fixed by the Einstein field equations. In alternative theories of gravity, though, different field equations can lead to different values of these parameters.

The combined results of the four classical tests can now be expressed as limits on the values of β and γ. As of 2017, these were consistent with the general relativistic predictions, with
$$\beta - 1 < 8 \times 10^{-5}, \qquad \gamma - 1 < 2 \times 10^{-5}. \tag{3.36}$$

The quantities β and γ are known as the Eddington-Robertson-Schiff parameters. The line element (3.35) can be further generalized to include off-diagonal terms and other forms of dependence on the mass distribution. Under a set of reasonable assumptions, one can obtain eight more PPN parameters, which label such things as preferred position and preferred frame effects. These parameters can be directly measured—in addition to the classical tests described in this chapter, they show up in tests of the equivalence principle, searches for violations of local Lorentz invariance, pulsar timing measurements, Lunar laser ranging, and so on. The PPN formalism thus provides a simple, compact way not only to test general relativity, but also to simultaneously compare it to a large class of alternative theories of gravity.

Further reading

For an overview of observational tests of general relativity, see Will's book [2] and review [4], and a review by Ni [13]. A long list of additional resources may be found in [14].

The Schwarzschild metric first appeared in [15]. For an instructive discussion of the apparent coordinate dependence of the metric and the way to extract physical predictions, see [12]. The exact solution of the geodesic equations for the Schwarzschild metric, given in terms of elliptic functions, can be found in [16]. More sophisticated approximations for the perihelion advance can be found in Section 15.3 of [8], Section 5.5 of [17], and Section 5.6.3 of [18].

The fascinating history of the problem of Mercury's anomalous perihelion advance is chronicled in a book by Roseveare [19]. An analysis based on measurements of the MESSENGER Mercury orbiter is given in [20]. The first measurements of the deflection of starlight by the Sun came from an expedition led by Eddington [21]. There has been a bit of controversy over the accuracy of those observations, but the original photographic plates were recently reanalyzed and the results confirmed; for an interesting discussion, see [22]. For a recent measurement of deflection of light by the Sun, based on over a million measurements of 541 celestial radio sources, see [23]. An introduction to gravitational lensing in cosmology may be found in [24]; for further resources, see [25].

The Shapiro time delay was introduced in [26]; for an extremely precise measurement using the Cassini spacecraft, see [27]. The first laboratory measurement of gravitational red shift was announced in [28]. For a measurement of gravitational time dilation over a height change of just .33 m, see [29]. For a nice description of relativistic effects in the GPS system, including the anecdote about the NTS-2 satellite, see [30].

The PPN formalism is discussed in much more detail in [2, 4, 13], and Chapter 39 of [11]. The limits quoted in eqn (3.36) are taken from [4].

4
Manifolds and tensors

John Wheeler famously summarized general relativity with the aphorism, "Spacetime tells matter how to move; matter tells spacetime how to curve" [31]. The first half of this saying, the way "spacetime tells matter how to move," describes the geodesic equation. It's now time to turn to the second half, the Einstein field equations, which specify the way "matter tells spacetime how to curve."

The curvature we are interested in is "intrinsic curvature," curvature that can be measured from within spacetime. When we think of the curvature of the Earth, we usually visualize a two-dimensional surface sitting in three-dimensional space. But we can also determine curvature from the surface itself: angles of triangles don't sum to π, regions enclosed by circles have areas different from πr^2, straight lines that start out parallel at the equator meet at the poles. Such intrinsic measurements are geometrical, and to grasp them we need to master the geometry of spacetime, differential geometry.

The next two chapters will build up this mathematics. Differential geometry has a reputation for being very hard. It's not, really; it's mainly just unfamiliar. The approach in this book will be mathematically correct, but informal, with some details relegated to Appendix A. We will start with the underlying setting, the manifold, and gradually build up structure—vectors and tensors in this chapter, differentiation, integration, and curvature in the next—until we have the ingredients to describe a general curved spacetime.

4.1 Manifolds

We start with a definition from mathematics. A function $f \colon M \to N$ between two spaces M and N is said to be a homeomorphism if it is continuous and has a continuous inverse. Intuitively, a homeomorphism can distort M by stretching or squeezing it, but can't cut or tear it. In settings in which derivatives are well-defined, a diffeomorphism is a homeomorphism in which f and f^{-1} are also differentiable.

A d-dimensional differentiable manifold M has several equivalent descriptions:

1. M looks locally like $\mathbb{R}^d$. That is, any point in M is contained in an open set $U \subset M$ that is homeomorphic to an open set $V \subset \mathbb{R}^d$. The word "local" is important here: M as a whole need not be homeomorphic to $\mathbb{R}^d$, but it can be covered by "patches" homeomorphic to subsets of $\mathbb{R}^d$. We need an additional rule to tell us how to "glue together" two overlapping patches. The formal definition is given in Section A.1 (see Fig. A.1), but the intuition is simple: given two overlapping patches with two different homeomorphisms to $\mathbb{R}^d$, the "transition functions" that relate the one homeomorphism to the other must be differentiable.

General Relativity: A Concise Introduction. Steven Carlip © Steven Carlip 2019.
Published in 2019 by Oxford University Press. DOI: 10.1093/oso/9780198822158.001.0001

> **Box 4.1 Diffeomorphism invariance**
>
> One of the central principles of general relativity is coordinate independence, or more formally, diffeomorphism invariance. Within a single coordinate patch, this is the statement that a transformation $x^\mu \to x^\mu + \xi^\mu$ should leave the physics unchanged. For the manifold as a whole, the statement is that a smooth enough mapping from M to itself—technically, a diffeomorphism—should not change the physics, as long as the mapping "drags" all physical and geometric objects along with it. As described in Section A.2, these two formulations are equivalent; they merely differ by a choice of whether to treat coordinate changes as "passive" or "active." As we shall see later, this invariance enormously restricts the form of the action, leads to conservation laws, and explains much of the structure of the Hamiltonian formulation.
>
> There is a slightly subtle technical issue, though. As Kretschmann argued soon after the publication of the Einstein field equations, any theory can be made diffeomorphism invariant by adding enough extra nondynamical structure. Modern statements of diffeomorphism invariance often add the condition of "background independence," and are formulated as a statement of invariance combined with an absence of "absolute" or "nondynamical" or "background" structures.

2. M can be consistently described in terms of local coordinates. To see this, pick a patch $U \subset M$ that is homeomorphic to an open set $V \subset \mathbb{R}^d$, and denote the homeomorphism by $x\colon U \to V$. For any point p in U, $x(p)$ is a point in $\mathbb{R}^d$, that is, an ordered set $(x^0(p), x^1(p), \ldots, x^{d-1}(p))$. The components $x^\mu(p)$ are then coordinates for p. The rule for overlaps is now simple: if a region can be described by two different coordinate systems, each set of coordinates must depend differentiably on the other.
3. M can be characterized by a set of "maps," in the ordinary English sense of the word. The image $x(U)$ of the open set U really is a map—a representation of U, perhaps distorted, in ordinary flat $\mathbb{R}^d$. As in ordinary map-making, it can be useful, and may sometimes be necessary, to describe M by a collection of overlapping maps. Much of the language of differential geometry is taken directly from map-making. The pair of open sets $(U \subset M, V \subset \mathbb{R}^d)$ and the homeomorphism x between them form a "chart," for instance, and a collection of charts that covers M is an "atlas."

These descriptions can easily be generalized to allow for boundaries, by replacing some of the open sets V with subsets of $\mathbb{R}^n$ with boundaries; see Section A.1.

Although the definition of a manifold requires a set of open sets and homeomorphisms, or equivalently a set of coordinates, these are not unique. Just as different collections of maps can describe the same region of the Earth, different charts can determine the same manifold. There are various ways to make this rigorous, described in Appendix A; technically, the statement is that two diffeomorphic manifolds are

Box 4.2 Manifolds and non-manifolds

It may be helpful to have some examples of manifolds and non-manifolds in mind. Here are a few illustrations in two dimensions (look at the surfaces, not the interiors):

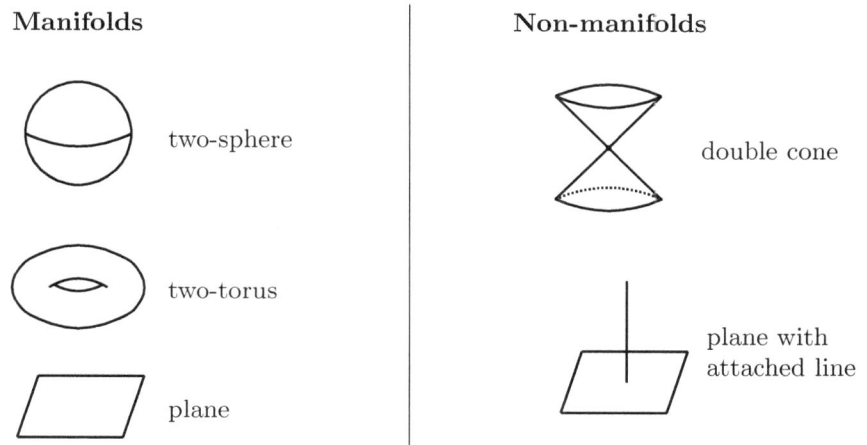

The non-manifolds shown here all have places where the dimension changes. Other more elaborate possibilities exist, but these examples are typical in physics.

considered identical. But the physical intuition, sometimes known as the "principle of general covariance," is simple: coordinates are human-made labels for the points in our spacetime manifold, and the Universe doesn't care what labels we use.

4.2 Tangent vectors

As a first step toward adding structure to a manifold, let us try to define vectors. Most of us first learn to picture a vector **v** as a little arrow with a magnitude and direction. Later, we learn to express **v** in terms of its components, reducing much of the geometry to algebra. But to write the components of a vector, we have to already know a set of basis vectors. In flat space, this is easy: basis vectors **i**, **j**, **k** are associated with any Cartesian coordinate system. But in a curved space, there are usually no preferred coordinates and no preferred basis. We need to somehow return to a more basis-independent description.

To do this, mathematicians have developed a clever trick. While the components of **v** depend on a basis, the directional derivative $\mathbf{v} \cdot \nabla$ along **v** is basis-independent. So we can simply *define* a vector—more specifically, a "tangent vector"—as a directional derivative. In particular, given a manifold M with local coordinates $\{x^\mu\}$ and

28 *Manifolds and tensors*

a parametrized curve $x^\mu(\sigma)$, the tangent vector v to the curve at a point p in the manifold is

$$v_p = \frac{d}{d\sigma}\bigg|_p = \frac{dx^\mu}{d\sigma}\frac{\partial}{\partial x^\mu}\bigg|_p. \tag{4.1}$$

The coefficient $dx^\mu/d\sigma$ is what we would normally call the tangent vector to the curve, but when combined this way with a partial derivative, the expression is coordinate-independent.

This definition determines a vector at a single point. An assignment of a vector to each point in M is called a vector field. Physicists usually treat these terms loosely, using "vector" and "vector field" interchangeably. The space of vectors at a point p is a d-dimensional vector space, the tangent space, denoted T_pM. The space of pairs (p, v_p), where v_p is a vector at p, is called the tangent bundle of M, denoted TM.

It is evident from (4.1) that in any given coordinate system, the vectors ∂_μ form a basis—any tangent vector can be written as a linear combination of these elements. Such a basis is called a coordinate basis. As always in linear algebra, we can form a new basis e_a from linear combinations:

$$e_a = e_a{}^\mu \partial_\mu \tag{4.2}$$

as long as the $d \times d$ matrix $e_a{}^\mu$ is invertible. (See Box 4.3 for notational conventions.)

In general, this new basis won't be a coordinate basis; that is, the e_a won't be derivatives with respect to new coordinates. There is a simple check. Since the e_a are differential operators, we can form the commutator $[e_a, e_b]$. In a coordinate basis, the commutator is zero, since partial derivatives of smooth enough functions commute. The converse also holds: if $[e_a, e_b] = 0$, one can find a local coordinate system around any point in which the e_a are derivatives with respect to the new coordinates. A coordinate basis is sometimes called "holonomic," and a non-coordinate basis "anholonomic."

This trick of expressing vectors as derivatives was introduced to give a basis-independent description, but it is also very useful for relating coordinate systems. In flat two-dimensional space, for instance, we can write the same vector in Cartesian and polar coordinates as

$$v = v^x \frac{\partial}{\partial x} + v^y \frac{\partial}{\partial y} = v^r \frac{\partial}{\partial r} + v^\theta \frac{\partial}{\partial \theta}. \tag{4.3}$$

This allows us to immediately read off the components in one coordinate system in terms of the other—for instance,

$$v^x = \left(v^x \frac{\partial}{\partial x} + v^y \frac{\partial}{\partial y}\right)x = \left(v^r \frac{\partial}{\partial r} + v^\theta \frac{\partial}{\partial \theta}\right)x = v^r \frac{\partial x}{\partial r} + v^\theta \frac{\partial x}{\partial \theta}. \tag{4.4}$$

This example illustrates the power of viewing a vector as a derivative: knowing the derivative means knowing the components, but more than that, it means knowing the components in any coordinate system. More generally, on a region of any manifold with two coordinate systems $\{x^\mu\}$ and $\{\bar{x}^\mu\}$,

$$v = v^\mu \frac{\partial}{\partial x^\mu} = \bar{v}^\nu \frac{\partial}{\partial \bar{x}^\nu} \Rightarrow v^\rho = \left(v^\mu \frac{\partial}{\partial x^\mu}\right)x^\rho = \bar{v}^\nu \frac{\partial x^\rho}{\partial \bar{x}^\nu}, \tag{4.5}$$

giving a simple way to see how components change under a change of coordinates.

Box 4.3 Conventions: part II

The choice of what letters to use as indices is not standardized, but this book will use a common convention. Lower case Greek letters, $\mu, \nu, \rho, \ldots$, will normally refer to some particular coordinate system. Lower case Roman letters from the beginning of the alphabet, $a, b, c, \ldots$, will refer to an arbitrary, not necessarily holonomic, basis. Sometimes Roman letters from the middle of the alphabet, $i, j, k, \ldots$, will refer to spatial indices in a spacetime; this will usually be clear from the context.

Tangent vectors will usually be denoted by lower case Roman letters $u, v, w, \ldots$, and cotangent vectors by lower case Greek letters $\alpha, \beta, \omega, \ldots$, with some exceptions. A basis e_a for tangent vectors is written with a lower index, while components v^a are written with an upper index. A basis for cotangent vectors e^a is written with an upper index, while components ω_a are written with a lower index. Recall also from (2.32) that ∂_μ means $\frac{\partial}{\partial x^\mu}$.

The objects $e_a{}^\mu$ of eqn (4.2) that define a general basis have competing names. From Greek, they are a dyad ("group of two," two dimensions), triad ("group of three," three dimensions), tetrad ("group of four," four dimensions), and so on. From German, they are a zweibein ("two legs," two dimensions), dreibein ("three legs," three dimensions), vierbein ("four legs," four dimensions), and so on, or vielbein ("many legs," any dimension). They can also be called a frame, from "reference frame," while the inverse $e^a{}_\mu$ can be called a coframe.

4.3 Cotangent vectors and gradients

The next task is to define a dot product, or inner product, of two vectors. In flat space, the dot product is a linear pairing of two vectors to give a real number. As in the preceding section, though, there are implicit assumptions that make this tricky to generalize: we need a concept of an orthonormal basis or a definition of magnitude and angle. For now, we will settle for a bit less, a linear pairing of a tangent vector with a *different* kind of object, a "dual vector" or "cotangent vector."

In linear algebra, any real vector space V has a dual space V^*, defined as the space of linear maps from V to $\mathbb{R}$. An element $\omega \in V^*$ is determined by its pairing (ω, v) with an arbitrary vector in V, where the pairing $(\bullet, \bullet)$ is linear in both arguments, that is,

$$(a\omega_1 + b\omega_2, v) = a(\omega_1, v) + b(\omega_2, v), \quad (\omega, a'v_1 + b'v_2) = a'(\omega, v_1) + b'(\omega, v_2) \quad (4.6)$$

for any real numbers a, b, a', b'. Given a basis e_a for V, the dual basis e^a for V^* (note the position of the index!) is defined by the pairing

$$(e^a, e_b) = \delta^a_b. \quad (4.7)$$

Any dual vector ω can be written as a linear combination $\omega = \omega_a e^a$, so

$$(\omega, v) = (\omega_a e^a, v^b e_b) = \omega_a v^b (e^a, e_b) = \omega_a v^a. \tag{4.8}$$

Perhaps the most familiar example of a dual space in physics comes from the Dirac notation in quantum mechanics, where a ket is a vector and a bra is a dual vector.

In differential geometry, a dual vector is called a cotangent vector, or, for reasons we'll see at the end of this chapter, a one-form. The space of dual vectors at a point p is the cotangent space T_p^*M; the space of pairs (p, ω_p), where ω_p is a cotangent vector at p, is the cotangent bundle T^*M. A cotangent vector may also be called a covariant vector, while a tangent vector is a contravariant vector.

The simplest example of a cotangent vector is the gradient of a function. Starting again with flat space, a gradient ∇f can be thought of as an object that determines a directional derivative along an arbitrary vector $\mathbf{v}$. In other words, the gradient can be defined by a pairing $(\nabla f, \mathbf{v}) = \nabla_\mathbf{v} \mathbf{f} = \mathbf{v} \cdot \nabla f$. Although $\mathbf{v}$ and ∇f are both vectors, geometrically they are quite different. A tangent vector is best visualized as an arrow tangent to a curve, while a gradient can be viewed as the set of contour lines of constant f; their inner product is, roughly, a count of the number of contour lines crossed by the arrow. This is a familiar idea to anyone who has gone hiking with a topographical map: the steepest rise, or largest directional derivative, is the direction in which the contour lines are closest together.

To generalize this idea to an arbitrary manifold M, let f be a real-valued function, $f\colon M \to \mathbb{R}$. The gradient of f, denoted df, is a cotangent vector defined by the pairing

$$(df, v) = v(f) = v^\mu \partial_\mu f, \tag{4.9}$$

where we use the fact that a tangent vector is now a directional derivative. The object df is not "infinitesimal," but the notation is deliberately parallel to the notation for differentials.

In a given coordinate system, the coordinates $\{x^\mu\}$ are themselves real-valued functions on M, at least in a coordinate patch, so they have gradients dx^μ. In fact, the dx^μ form a dual basis to the coordinate basis ∂_μ:

$$(dx^\mu, \partial_\nu) = \left(\frac{\partial}{\partial x^\nu}\right) x^\mu = \delta^\mu_\nu. \tag{4.10}$$

In this basis, any gradient df has components

$$df = \partial_\mu f \, dx^\mu, \tag{4.11}$$

as can be checked from the definitions:

$$(df, \partial_\nu) = \partial_\nu f = (\partial_\mu f)\delta^\mu_\nu = \partial_\mu f (dx^\mu, \partial_\nu) = (\partial_\mu f \, dx^\mu, \partial_\nu),$$

where the last equality comes from the linearity of the pairing $(\bullet, \bullet)$. Just as in the case of tangent vectors, we can form a more general basis

$$e^a = e^a{}_\mu dx^\mu. \tag{4.12}$$

It is easy to check that the basis (4.12) will be dual to (4.2) provided that

$$e_a{}^\mu e^b{}_\mu = \delta^a_b \Leftrightarrow e_a{}^\mu e^a{}_\nu = \delta^\mu_\nu. \qquad (4.13)$$

As with tangent vectors, the expression of a cotangent vector in terms of a coordinate basis makes it easy to see how the components transform under a coordinate change. The analog of (4.5) is now

$$\omega = \omega_\mu dx^\mu = \bar{\omega}_\nu d\bar{x}^\nu \Rightarrow \omega_\rho = \left(\omega_\mu dx^\mu, \frac{\partial}{\partial x^\rho}\right) = \left(\bar{\omega}_\nu d\bar{x}^\nu, \frac{\partial}{\partial x^\rho}\right) = \bar{\omega}_\nu \frac{\partial \bar{x}^\nu}{\partial x^\rho}. \qquad (4.14)$$

Note that the positions of x and $\bar{x}$ are reversed when we switch from tangent vectors (4.5) to cotangent vectors (4.14), so $(\omega, v) = \omega_\mu v^\mu = \bar{\omega}_\mu \bar{v}^\mu$, as required.

4.4 Tensors

We have defined a cotangent vector ω at a point p as a linear function that takes as its argument a tangent vector and gives back a real number. That is,

$$\omega: T_p M \to \mathbb{R} \quad \text{with } v \mapsto (\omega, v). \qquad (4.15)$$

We could equally well have begun with gradients and cotangent vectors, and defined a tangent vector as a linear function that takes cotangent vectors to real numbers. Viewed this way, there is an obvious generalization. A type (k, ℓ) tensor T at a point p is a multilinear map that takes as its argument k cotangent vectors and ℓ tangent vectors and gives a real number. That is,

$$T: \underbrace{T_p^* M \times \cdots \times T_p^* M}_{k \text{ times}} \times \underbrace{T_p M \times \cdots \times T_p M}_{\ell \text{ times}} \to \mathbb{R}, \qquad (4.16)$$

where T is linear in each of its arguments. Thanks to this linearity, a tensor is completely determined by its action on a set of basis vectors: given a basis e_b for tangent vectors and its dual basis e^a for cotangent vectors, the components

$$T(e^{a_1}, e^{a_2}, \ldots, e^{a_k}, e_{b_1}, e_{b_2}, \ldots, e_{b_\ell}) = T^{a_1 a_2 \ldots a_k}{}_{b_1 b_2 \ldots b_\ell} \qquad (4.17)$$

completely determine T. The total number of indices, $k + \ell$, is called the rank of T. As before, a tensor field is the assignment of a tensor to each point in M.

A cotangent vector is a $(0, 1)$ tensor, with components $\omega(e_a) = \omega_a$. A tangent vector is a $(1, 0)$ tensor, with components $v(e^a) = v^a$. A function, or scalar, is a $(0, 0)$ tensor. The Kronecker delta is a $(1, 1)$ tensor—that is, there is a tensor δ whose components are

$$\delta(e^a, e_b) = \delta^a_b. \qquad (4.18)$$

For δ to be a genuine tensor, this equation must be independent of the choice of basis; that is, if (4.18) holds for one basis it must hold for any other. This is not hard to verify.

Box 4.4 Symmetrization

A general tensor has no particular symmetries, but it is often useful to form symmetric and antisymmetric combinations. Start with a type $(0, p)$ tensor—the generalization is obvious—with components $w_{a_1 a_2 \ldots a_p}$. Let π denote a permutation of the indices, $(a_1, a_2, \ldots, a_p) \to (\pi(a_1), \pi(a_2), \ldots, \pi(a_p))$. The symmetrization of w, denoted by parentheses around the indices, is given by a sum over all permutations,

$$w_{(a_1 a_2 \ldots a_p)} = \frac{1}{p!} \sum_\pi w_{\pi(a_1)\pi(a_2)\ldots\pi(a_p)},$$

while the antisymmetrization, denoted by square brackets, is

$$w_{[a_1 a_2 \ldots a_p]} = \frac{1}{p!} \sum_\pi (-1)^\pi w_{\pi(a_1)\pi(a_2)\ldots\pi(a_p)},$$

where $(-1)^\pi$ is $+1$ if π is an even permutation and -1 if it is odd. For example,

$$w_{[abc]} = \frac{1}{6}\left(w_{abc} + w_{bca} + w_{cab} - w_{bac} - w_{acb} - w_{cba}\right).$$

On the other hand, not every object with indices is a tensor. For instance, the Levi-Civita alternating symbol

$$\tilde{\epsilon}_{a_0 a_1 \ldots a_{d-1}} = \begin{cases} +1 & \text{if } (a_0, a_1, \ldots) \text{ is an even permutation of } (0, 1, \ldots, d-1) \\ -1 & \text{if } (a_0, a_1, \ldots) \text{ is an odd permutation of } (0, 1, \ldots, d-1) \\ 0 & \text{otherwise} \end{cases} \quad (4.19)$$

is not a tensor: while we can define a tensor T for which $T(e_{a_0}, e_{a_1}, \ldots, e_{a_{d-1}}) = \tilde{\epsilon}_{a_0 a_1 \ldots a_{d-1}}$ in *some* basis, the equality will no longer hold in other bases.

Given a basis for vectors, it is useful to build a corresponding basis for arbitrary tensors. It's tempting to try an ordered set $(e_{a_1}, e_{a_2}, \ldots, e_{a_k}, e^{b_1}, e^{b_2}, \ldots, e^{b_\ell})$. This doesn't quite work, though. To see the problem, consider two tangent vectors v and w, and a tensor with components are $T^{ab} = v^a w^b$. If we try to write T^{ab} in terms of a basis (e_a, e_b), there's an ambiguity:

$$T = (v^a e_a)(w^b e_b) = (c\, v^a e_a)(c^{-1} w^b e_b) \quad \text{for any constant } c \neq 0.$$

This makes it unclear exactly what number goes with what basis element.

But this is obviously just a technical annoyance. To remove the ambiguity, mathematicians define the tensor product $\otimes$ as an equivalence class of ordered pairs:

$$e \otimes f = (e, f)/\sim \quad \text{with } (ce, f) \sim (e, cf) \sim c(e, f), \quad c \in \mathbb{R}. \quad (4.20)$$

This simply means we allow constants to freely shuffle back and forth among basis elements. Then $e_{a_1} \otimes e_{a_2} \otimes \cdots \otimes e_{a_k} \otimes e^{b_1} \otimes e^{b_2} \otimes \cdots \otimes e^{b_\ell}$ is a basis for (k, ℓ) tensors,

$$T = T^{a_1 a_2 \ldots a_k}{}_{b_1 b_2 \ldots b_\ell}\, e_{a_1} \otimes e_{a_2} \otimes \cdots \otimes e_{a_k} \otimes e^{b_1} \otimes e^{b_2} \otimes \cdots \otimes e^{b_\ell}. \tag{4.21}$$

As before, this makes it easy to track changes of components under a change of coordinates. For a $(1,1)$ tensor, for instance,

$$T = T^\mu{}_\nu \frac{\partial}{\partial x^\mu} \otimes dx^\nu = \bar{T}^\lambda{}_\sigma \frac{\partial}{\partial \bar{x}^\lambda} \otimes d\bar{x}^\sigma \;\;\Rightarrow\;\; T^\rho{}_\tau = \bar{T}^\lambda{}_\sigma \frac{\partial x^\rho}{\partial \bar{x}^\lambda} \frac{\partial \bar{x}^\sigma}{\partial x^\tau}, \tag{4.22}$$

the natural generalization of (4.5) and (4.14).

We will need one more tensor operation. Contraction is a process that forms a type $(k-1, \ell-1)$ tensor from a type (k, ℓ) tensor by a generalized trace. In terms of components, we set an upper index equal to a lower index and sum:

$$S^{a_1 a_2 \ldots \hat{a}_m \ldots a_k}{}_{b_1 b_2 \ldots \hat{b}_n \ldots b_\ell} = T^{a_1 a_2 \ldots c \ldots a_k}{}_{b_1 b_2 \ldots c \ldots b_\ell}, \tag{4.23}$$

where a hat over an index means the index has been omitted. In terms of the tensor itself,

$$S = T(\ldots, e^c, \ldots, e_c, \ldots), \tag{4.24}$$

that is, we plug a basis element e^c into the mth cotangent vector slot and a basis element e_c into the nth tangent vector slot and sum over the basis. It may be checked that the result doesn't depend on the choice of basis, as long as $\{e^a\}$ and $\{e_a\}$ are dual bases. Contractions on different indices, on the other hand, will typically give us different $(k-1, \ell-1)$ tensors.

While it may be obvious, it is worth emphasizing that tensors offer a simple way to write down diffeomorphism-invariant equations (see Box 4.1). While the *components* of a tensor may depend on a choice of coordinate basis, the tensor itself does not, and a tensor equation of the form $T_1 = T_2$ is equally valid in any coordinate system.

4.5 The metric

The metric, which we first encountered in Chapter 2, can now be given a rigorous definition. A metric on a manifold M is a symmetric nondegenerate type $(0,2)$ tensor field, $g = g_{ab} e^a \otimes e^b$, where "nondegenerate" means that the eigenvalues of the matrix of components g_{ab} are all nonzero. In a coordinate basis,

$$g = g_{\mu\nu} dx^\mu \otimes dx^\nu, \tag{4.25}$$

a mathematically well-defined version of (2.28). The inverse metric, a type $(2,0)$ tensor $g^{ab} e_a \otimes e_b$, is defined as in Chapter 2: its components in any basis are given by the matrix inverse of the components of the metric,

$$g_{ab} g^{bc} = \delta_a^c. \tag{4.26}$$

It must be checked that g^{ab} is really a tensor, that is, if (4.26) holds in one basis, then it holds in any other. This is indeed the case.

The metric provides an inner product between a pair of vectors, $g(u, v) = g_{ab} u^a v^b$, although the result is only positive definite if the signature of the metric is Riemannian

34 *Manifolds and tensors*

(see Section 2.3). The metric can also be used to "raise and lower indices," that is, to map between tangent and cotangent spaces. Suppose we start with a tangent vector u. Then the quantity $g(u, \bullet)$—the metric with one empty slot—is a cotangent vector: it pairs with another vector v to give a real number $g(u, v)$. The tangent vector u and the associated cotangent vector $g(u, \bullet)$ are conventionally denoted by the same letter, distinguished by the index position. In terms of the components in an arbitrary basis,

$$u_a = g_{ab} u^b, \quad u^c = g^{cd} u_d. \tag{4.27}$$

Indices of higher rank tensors can be raised and lowered in the same way.

We will later need a useful identity relating derivatives of the metric and the inverse metric. Let δ represent any derivative—a partial derivative or, later, a functional variation. Using the fact that $\delta(g_{ab} g^{bc}) = 0$, it is easy to check that

$$\delta g^{ab} = -g^{ac} g^{bd} \delta g_{cd}. \tag{4.28}$$

We will also need one more object built out of the metric. In a given basis e^a, the components g_{ab} form a $d \times d$ matrix. The determinant of this matrix,

$$g = \det |g_{ab}|, \tag{4.29}$$

is also denoted g, a confusing notation but one that is usually clear in context. This determinant is not a tensor—it is basis-dependent—but it will be indispensable in our treatment of integration in the next chapter.

Of course, a manifold can have many symmetric $(0, 2)$ tensors. Choosing one to be the metric is a statement about the geometry: the metric determines the geodesics, and, as we will see later, the curvature. For physics, the main purpose of the Einstein field equations will be to determine just which $(0, 2)$ tensor is the physical metric.

4.6 Isometries and Killing vectors

As elsewhere in physics, it is often useful to consider settings with special symmetries. The Schwarzschild metric, for instance, is symmetric under rotations and time translations. The real world has no such exact symmetries, of course, but approximate symmetries can often vastly simplify calculations.

A symmetry of the metric tensor is called an isometry. To determine the isometries requires a bit of care, since a change of position affects not only the components of the metric, but also the basis. Consider a shift $x^\mu \to x^\mu + \xi^\mu$, where ξ^μ is small. The cotangent basis vectors and the components $g_{\mu\nu}$ in (4.25) will both change:

$$dx^\mu \to dx^\mu + d\xi^\mu = dx^\mu + \partial_\rho \xi^\mu \, dx^\rho + \mathcal{O}(\xi^2),$$
$$g_{\mu\nu}(x) \to g_{\mu\nu}(x + \xi) = g_{\mu\nu}(x) + \xi^\rho \partial_\rho g_{\mu\nu} + \mathcal{O}(\xi^2). \tag{4.30}$$

Hence, after changing some dummy indices,

$$g_{\mu\nu} dx^\mu \otimes dx^\nu \to \left(g_{\mu\nu} + g_{\mu\rho} \partial_\nu \xi^\rho + g_{\rho\nu} \partial_\mu \xi^\rho + \xi^\rho \partial_\rho g_{\mu\nu} + \mathcal{O}(\xi^2) \right) dx^\mu \otimes dx^\nu. \tag{4.31}$$

The shift $x \to x + \xi$ is an isometry if g, as a tensor, doesn't change:

$$g_{\mu\rho} \partial_\nu \xi^\rho + g_{\rho\nu} \partial_\mu \xi^\rho + \xi^\rho \partial_\rho g_{\mu\nu} = 0. \tag{4.32}$$

Equation (4.32) is called the Killing equation, after the mathematician Wilhelm Killing, and its solutions are Killing vectors. In four dimensions, the Killing equation is actually ten equations for four unknowns, a badly overdetermined system. Generically there are no solutions; only very special metrics admit isometries.

It will be useful later to write the equation in a slightly different form. Using the metric to lower an index, it may be checked that the Killing equation becomes

$$\partial_\mu \xi_\nu + \partial_\nu \xi_\mu - 2\Gamma^\rho_{\mu\nu} \xi_\rho = 0, \tag{4.33}$$

where $\Gamma^\rho_{\mu\nu}$ are the components of the Levi-Civita connection given in eqn (2.40). As we shall see in the next chapter, the left-hand side is a combination of covariant derivatives, $\nabla_\mu \xi_\nu + \nabla_\nu \xi_\mu$.

4.7 Orthonormal bases

A basis e_a for tangent vectors is a collection $\{e_0, e_1, \ldots, e_{d-1}\}$ of d vectors. Now that we have a metric, we can define an inner product $g(e_a, e_b) = g_{ab}$ of these vectors. A basis is orthogonal if the off-diagonal elements of g_{ab} are zero; it is orthonormal if, in addition, the diagonal elements are all ± 1:

$$g_{ab} = \eta_{ab} = \begin{cases} \pm 1 & \text{if } a = b \\ 0 & \text{if } a \neq b \end{cases}. \tag{4.34}$$

The number of $+1$ and -1 entries depends on the signature of the metric. For a Riemannian metric, all are $+1$; for a Lorentzian metric with the West Coast sign convention (Box 2.2), one entry is $+1$ and the rest are -1. The dual basis of an orthonormal basis is also orthonormal:

$$g^{ab} = \eta^{ab} = \begin{cases} \pm 1 & \text{if } a = b \\ 0 & \text{if } a \neq b \end{cases} \quad \text{with } \eta_{ab}\eta^{bc} = \delta^c_a. \tag{4.35}$$

The vectors that constitute an orthonormal basis, like any vectors, can be expressed in terms of a coordinate basis, $e_a = e_a{}^\mu \partial_\mu$. Orthonormality then implies that $g(e_a, e_b) = \eta_{ab} = e_a{}^\mu e_b{}^\nu g(\partial_\mu, \partial_\nu)$, or

$$g_{\mu\nu} e_a{}^\mu e_b{}^\nu = \eta_{ab} \iff \eta^{ab} e_a{}^\mu e_b{}^\nu = g^{\mu\nu}. \tag{4.36}$$

The relationship (4.13) between a basis and its dual basis now means

$$e^a{}_\mu = \eta^{ab} g_{\mu\nu} e_b{}^\nu, \tag{4.37}$$

so we can again raise and lower indices with appropriate components of the metric.

In an orthonormal basis, the metric is simply

$$g = \pm e^0 \otimes e^0 \pm e^1 \otimes e^1 \pm \cdots \pm e^{d-1} \otimes e^{d-1}. \tag{4.38}$$

If the line element is diagonal, this makes it trivial to write down an orthonormal basis for cotangent vectors. For instance, for the Schwarzschild metric (3.1),

$$e^0 = \left(1 - \frac{2m}{r}\right)^{1/2} dt, \quad e^1 = \left(1 - \frac{2m}{r}\right)^{-1/2} dr, \quad e^2 = r\, d\theta, \quad e^3 = r \sin\theta\, d\varphi. \tag{4.39}$$

For nondiagonal line elements, the same thing can be accomplished by completing the square, writing ds^2 as a sum or difference of squares.

An orthonormal basis is not unique, but bases are related by simple transformations, those that preserve η_{ab}—rotations for Riemannian signature, Lorentz transformations for Lorentzian signature. An orthonormal basis may be viewed as a local reference frame; invariance under a basis change is then "local Lorentz invariance."

4.8 Differential forms

One particular class of tensors will prove useful later. A differential p-form (or p-form for short) is a totally antisymmetric type $(0,p)$ tensor. A zero-form is a function; a one-form is a cotangent vector; a two-form is an antisymmetric $(0,2)$ tensor, that is, a tensor with components $\alpha_{ab} = -\alpha_{ba}$. Because of the complete antisymmetry, a d-form on a d-dimensional manifold has only one independent component, and can always be written as $\alpha_{a_0 a_1 \ldots a_{d-1}} = \hat{\alpha} \tilde{\epsilon}_{a_0 a_1 \ldots a_{d-1}}$, where $\tilde{\epsilon}$ is the Levi-Civita symbol (4.19).

The product of two forms is not itself a form—$\alpha_a \beta_b$ is not antisymmetric, for instance—but it can be made into a form by antisymmetrizing. Using the notation of Box 4.4, we define the wedge product of a p-form α and a q-form β to be the $(p+q)$-form

$$(\alpha \wedge \beta)_{a_1 a_2 \ldots a_{p+q}} = \frac{(p+q)!}{p!q!} \alpha_{[a_1 \ldots a_p} \beta_{a_{p+1} \ldots a_{p+q}]}, \qquad (4.40)$$

antisymmetrized on all $p+q$ indices. It is easy to check that

$$\alpha \wedge \beta = (-1)^{pq} \beta \wedge \alpha. \qquad (4.41)$$

The normalization factors are chosen to make some equations simple: for instance, if α and β are both one-forms, $(\alpha \wedge \beta)_{ab} = \alpha_a \beta_b - \alpha_b \beta_a$. As usual, though, one must keep careful track.

The basis vectors $e^{a_1} \wedge e^{a_2} \wedge \cdots \wedge e^{a_p}$, or in a coordinate basis, $dx^{\mu_1} \wedge dx^{\mu_2} \wedge \cdots \wedge dx^{\mu_p}$, are a basis for the p-forms. As we will see in the next chapter, differentiation of p-forms—"Cartan calculus"—is particularly simple. Section A.5 shows that certain integral identities are also most easily expressed in the language of forms. An additional operation, the Hodge dual, is described in Section A.5 of Appendix A; in a d-dimensional spacetime, it allows us to map a p-form to a $(d-p)$-form.

Further reading

For this and the next chapter, some of the mathematical details have been moved to Appendix A. Two very good introductions to differential geometry are Isham's *Modern Differential Geometry for Physicists* [32] and Choquet-Bruhat and DeWitt-Morette's *Analysis, Manifolds and Physics* [33]. A beautiful, intuitive introduction to the topology of manifolds is Weeks' *The Shape of Space* [34]. Readers might also look at other textbooks on general relativity: for instance, [11] and [35] both have more detailed and technical introductions to the mathematics. For differential forms and Cartan calculus, a book by Henri Cartan, the son of the inventor, is somewhat formal but can be quite interesting [36].

Kretschmann's argument about diffeomorphism invariance, discussed in Box 4.1, appeared in [37]. Philosophers of physics still debate this issue; a nice physicist's treatment is [38].

5
Derivatives and curvature

We now have a collection of objects that live on a manifold. The next step is to develop calculus—differentiation and integration—for these objects.

5.1 Integration

Let us start with integration. Consider a d-dimensional manifold M and a coordinate patch U with coordinates $\{x^\mu\}$. We would like to make sense of the expression

$$I = \int_U \mathcal{F}(x)\, d^d x. \tag{5.1}$$

What kind of object can the integrand $\mathcal{F}$ be?

Our basic criterion will be that the integral must not depend on a choice of coordinates. This is not just a philosophical preference: a manifold often requires more than one coordinate patch, and if the integral (5.1) doesn't have a consistent value on the overlaps, there is no way to add up the contributions of different patches.

Consider a new set of coordinates $\{\bar{x}^\mu\}$. We know from ordinary calculus that under a coordinate change, the integration measure changes by a Jacobian, the determinant of the matrix of derivatives $\partial \bar{x}^\mu / \partial x^\nu$:

$$d^d \bar{x} = J\, d^d x \quad \text{with } J = \det\left|\frac{\partial \bar{x}}{\partial x}\right|, \tag{5.2}$$

We thus need

$$\bar{\mathcal{F}} = J^{-1} \mathcal{F} = \mathcal{F} \det\left|\frac{\partial x}{\partial \bar{x}}\right|. \tag{5.3}$$

An object that transforms like (5.3) is called a scalar density of weight -1. If we have a metric, we automatically have one such object. Under a coordinate change, the components of the metric transform, in analogy with (4.14), as

$$\bar{g}_{\mu\nu} = g_{\rho\sigma} \frac{\partial x^\rho}{\partial \bar{x}^\mu} \frac{\partial x^\sigma}{\partial \bar{x}^\nu}. \tag{5.4}$$

As in Section 4.5, denote by g the determinant of the components of the metric in coordinates $\{x^\mu\}$, and by $\bar{g}$ the determinant in coordinates $\{\bar{x}^\mu\}$. Treating (5.4) as a matrix equation and taking the determinant, we see that

$$\bar{g} = g \det\left|\frac{\partial x}{\partial \bar{x}}\right|^2, \tag{5.5}$$

General Relativity: A Concise Introduction. Steven Carlip © Steven Carlip 2019.
Published in 2019 by Oxford University Press. DOI: 10.1093/oso/9780198822158.001.0001

38 Derivatives and curvature

since the determinant of a product is equal to the product of the determinants. Hence if f is a scalar, the combination
$$\mathcal{F} = \sqrt{|g|}\, f \tag{5.6}$$
is a scalar density of weight -1, and can be integrated. The integration measure $\sqrt{|g|}\, d^4x$ is sometimes called the "volume element."

(The absolute value in $\sqrt{|g|}$ is needed because, depending on the metric signature, g can be positive or negative. For a four-dimensional Lorentzian metric, g is negative, so many authors write $\sqrt{-g}$. As discussed in Section A.5, even in the absence of a metric, we can still define an integral on a manifold; the object that can be integrated is then a d-form.)

5.2 Why derivatives are more complicated

Our next task is to define derivatives of tensor fields. The obvious first guess would be to simply take derivatives of the components of a tensor. This doesn't work. As a simple illustration, consider a cotangent vector with components ω_ν. For $\partial_\mu \omega_\nu$ to be the components of a tensor, we would need

$$\left(\frac{\partial \omega_\nu}{\partial x^\mu}\right) dx^\mu \otimes dx^\nu \stackrel{?}{=} \left(\frac{\partial \bar{\omega}_\sigma}{\partial \bar{x}^\rho}\right) d\bar{x}^\rho \otimes d\bar{x}^\sigma$$

under a basis change. But this is *not* the case. Using the transformation (4.14) and the usual chain rule, we have

$$\left(\frac{\partial \omega_\nu}{\partial x^\mu}\right) = \frac{\partial \bar{x}^\rho}{\partial x^\mu} \frac{\partial}{\partial \bar{x}^\rho}\left(\bar{\omega}_\sigma \frac{\partial \bar{x}^\sigma}{\partial x^\nu}\right) = \frac{\partial \bar{\omega}_\sigma}{\partial \bar{x}^\rho} \frac{\partial \bar{x}^\rho}{\partial x^\mu} \frac{\partial \bar{x}^\sigma}{\partial x^\nu} + \bar{\omega}_\sigma \frac{\partial^2 \bar{x}^\sigma}{\partial x^\mu \partial x^\nu}. \tag{5.7}$$

The second term is a disaster, completely spoiling the tensorial properties.

With a little thought, it's clear where we've gone wrong. We want the derivative of the cotangent vector $\omega_\mu dx^\mu$, not just of its components in some basis. If the basis vectors dx^μ aren't constant, that means we need their derivatives as well. But this has led us in a circle: we can perhaps define the derivative of a vector, but only if we already know the derivatives of d basis vectors.

5.3 Derivatives of p-forms: Cartan calculus

Before tackling the general case, let us look at a special simple instance. If we take the antisymmetric combination of derivatives in (5.7), $\partial_\mu \omega_\nu - \partial_\nu \omega_\mu$, the undesirable non-tensorial term cancels out. It's not hard to check that the same is true for the completely antisymmetrized derivative of any p-form. Define the exterior derivative d acting on a p-form α as the operation

$$(d\alpha)_{\mu_1 \mu_2 \ldots \mu_{p+1}} = (p+1)\partial_{[\mu_1} \alpha_{\mu_2 \ldots \mu_{p+1}]}. \tag{5.8}$$

With this special combination of derivatives, the quantity

$$(d\alpha)_{\mu_1 \mu_2 \ldots \mu_{p+1}} dx^{\mu_1} \otimes dx^{\mu_2} \otimes \cdots \otimes dx^{\mu_{p+1}}$$

is a genuine tensor—specifically, an antisymmetric type $(0, p+1)$ tensor, or a $(p+1)$-form. If α is a zero-form, a scalar, then d is just the gradient defined in Section 4.3. If α is a one-form, the normalization is chosen so that

$$(d\alpha)_{\mu\nu} = \partial_\mu \alpha_\nu - \partial_\nu \alpha_\mu.$$

If α is a p-form and β is a q-form, the exterior derivative obeys a version of the usual product rule,

$$d(\alpha \wedge \beta) = (d\alpha) \wedge \beta + (-1)^p \alpha \wedge (d\beta). \tag{5.9}$$

It is easy to see that $d^2\alpha = 0$ for any form α: the antisymmetrization in (5.8) means that every term occurs in a pair $(\partial_\mu \partial_\nu - \partial_\nu \partial_\mu)\alpha_{\rho...}$, which vanishes because partial derivatives commute. This allows us to give an alternative definition of the exterior derivative, one that is very useful for calculation. The exterior derivative d is determined by three features:

1. For any function f, df is the gradient, $df = \partial_\mu f dx^\mu$;
2. d obeys the product rule (5.9);
3. $d^2 = 0$.

We can apply this, for example, to a one-form ω:

$$\begin{aligned}d\omega &= d(\omega_\nu dx^\nu) = d\omega_\nu \wedge dx^\nu + \omega_\nu d(dx^\nu) = (\partial_\mu \omega_\nu) dx^\mu \wedge dx^\nu + 0 \\ &= (\partial_\mu \omega_\nu)(dx^\mu \otimes dx^\nu - dx^\nu \otimes dx^\mu) = (\partial_\mu \omega_\nu - \partial_\nu \omega_\mu) dx^\mu \otimes dx^\nu,\end{aligned} \tag{5.10}$$

where the normalization (4.40) of the wedge product has been used.

The calculus of p-forms is called "Cartan calculus" after its inventor, Élie Cartan. In addition to its usefulness for computation, it is valuable for manipulating integrals and analyzing topological features of manifolds. Appendix A includes a brief discussion of its role in Stokes' theorem and de Rham cohomology.

5.4 Connections and covariant derivatives

Calculus is easy for differential forms, but this is not enough. The metric, for instance, is not a differential form—it's not antisymmetric—but we still need its derivatives. This will require a new structure on our manifold M, a "connection."

As noted earlier, what we need are the derivatives of basis vectors. The derivative of a basis vector e_a must itself be a vector, and thus a linear combination of basis vectors,

$$\partial_\mu e_a = \Gamma_\mu{}^b{}_a e_b \tag{5.11}$$

for some coefficients $\Gamma_\mu{}^b{}_a$. These coefficients are called the components of the connection, because they allow us to "connect" the basis vectors (or tangent spaces) at nearby points. Equation (5.11) is expressed in a mixed basis: derivatives are written in a coordinate basis ∂_μ, while vectors are written in a general basis e_a. This may be slightly confusing at first, but it's actually helpful in keeping track of indices.

Box 5.1 Conventions: part III

This book uses ∂_μ to denote a coordinate derivative, and ∇_μ to denote a covariant derivative. In some books and papers, especially older ones, a partial derivative may also be denoted by a comma or a single bar: $\partial_\rho A_\mu = A_{\mu,\rho} = A_{\mu|\rho}$. A covariant derivative may be denoted by a semicolon or a double bar: $\nabla_\rho A_\mu = A_{\mu;\rho} = A_{\mu||\rho}$.

Assuming derivatives obey the usual product rule (or "Leibniz rule"), we have

$$\partial_\mu(v^a e_a) = (\partial_\mu v^a) e_a + v^a \Gamma_\mu{}^b{}_a e_b = \left(\partial_\mu v^a + \Gamma_\mu{}^a{}_b v^b\right) e_a. \tag{5.12}$$

The coefficient of e_a in the last term is called the covariant derivative of v (strictly speaking, the components of the covariant derivative in the chosen basis), and is denoted

$$\nabla_\mu v^a = \partial_\mu v^a + \Gamma_\mu{}^a{}_b v^b. \tag{5.13}$$

So far, the components of the connection have been essentially arbitrary. In the next section, we will add conditions that determine the connection in terms of the metric. But we can already make a few general observations.

- **Covariant derivatives of other tensors**

  A scalar f contains no basis elements, so its covariant derivative is just its ordinary derivative. If we again postulate a product rule, we can obtain covariant derivatives of other tensors. For instance, let ω_a be a cotangent vector, and contract it with a tangent vector v^a to form a scalar. Then $\nabla_\mu(\omega_a v^a) = \partial_\mu(\omega_a v^a)$, so

$$\nabla_\mu(\omega_a v^a) - \partial_\mu(\omega_a v^a) = 0$$
$$= (\nabla_\mu \omega_a) v^a + \omega_a \left(\partial_\mu v^a + \Gamma_\mu{}^a{}_b v^b\right) - (\partial_\mu \omega_a) v^a - \omega_a(\partial_\mu v^a)$$
$$= \left(\nabla_\mu \omega_a - \partial_\mu \omega_a + \Gamma_\mu{}^b{}_a \omega_b\right) v^a.$$

Since this must hold for any tangent vector v^a, we can read off $\nabla_\mu \omega_a$:

$$\nabla_\mu \omega_a = \partial_\mu \omega_a - \Gamma_\mu{}^b{}_a \omega_b. \tag{5.14}$$

This trick extends to arbitrary tensors:

$$\nabla_\mu T^{a_1 a_2 \ldots a_k}{}_{b_1 b_2 \ldots b_\ell} = \partial_\mu T^{a_1 a_2 \ldots a_k}{}_{b_1 b_2 \ldots b_\ell}$$
$$+ \Gamma_\mu{}^{a_1}{}_c T^{c a_2 \ldots a_k}{}_{b_1 b_2 \ldots b_\ell} + \Gamma_\mu{}^{a_2}{}_c T^{a_1 c \ldots a_k}{}_{b_1 b_2 \ldots b_\ell} + \ldots$$
$$- \Gamma_\mu{}^c{}_{b_1} T^{a_1 a_2 \ldots a_k}{}_{c b_2 \ldots b_\ell} - \Gamma_\mu{}^c{}_{b_2} T^{a_1 a_2 \ldots a_k}{}_{b_1 c \ldots b_\ell} - \ldots \tag{5.15}$$

with a connection term occurring with a $+$ sign for each upper index and a $-$ sign for each lower index.

- **Derivatives of dual basis vectors**

  We can obtain the same result by using (5.11) to find the derivatives of the dual basis. We again need a product rule, this time for the pairing (4.7):

  $$\partial_\mu (e^a, e_b) = \partial_\mu \delta^a_b = 0$$
  $$= (\partial_\mu e^a, e_b) + (e^a, \partial_\mu e_b) = (\partial_\mu e^a, e_b) + (e^a, \Gamma_\mu{}^c{}_b e_c) = (\partial_\mu e^a, e_b) + \Gamma_\mu{}^a{}_b,$$

  from which we can read off

  $$\partial_\mu e^a = -\Gamma_\mu{}^a{}_b e^b. \tag{5.16}$$

  If we expand an arbitrary tensor in a basis, as in (4.21), and differentiate using (5.11) and (5.16), we recover the covariant derivative (5.15).

- **Change of basis**

  We defined our connection coefficients in a basis. Consider a new basis $\bar{e}_a$, with

  $$e_a = \Lambda_a{}^c \bar{e}_c. \tag{5.17}$$

  Substituting into (5.11), we have

  $$\partial_\mu (\Lambda_a{}^c \bar{e}_c) = \Gamma_\mu{}^b{}_a e_b = \Gamma_\mu{}^b{}_a (\Lambda_b{}^c \bar{e}_c)$$
  $$= (\partial_\mu \Lambda_a{}^c) \bar{e}_c + \Lambda_a{}^c \partial_\mu \bar{e}_c = (\partial_\mu \Lambda_a{}^c) \bar{e}_c + \Lambda_a{}^d \bar{\Gamma}_\mu{}^c{}_d \bar{e}_c,$$

  from which we see that

  $$\Gamma_\mu{}^b{}_a = (\partial_\mu \Lambda_a{}^c) \Lambda^{-1}{}^b{}_c + \Lambda_a{}^d \bar{\Gamma}_\mu{}^c{}_d \Lambda^{-1}{}^b{}_c. \tag{5.18}$$

  This expression may look familiar to readers who have seen gauge theories: it is essentially the same as the gauge transformation of a non-Abelian gauge field. Note that while a connection does not transform as a tensor, the *difference* between two connections does: the $(\partial \Lambda) \Lambda^{-1}$ term in (5.18) cancels out in such a difference.

  An important instance of (5.18) occurs when $\bar{e}$ is our coordinate basis. Equation (5.17) is then just $e_a = e_a{}^\mu \partial_\mu$, and, recalling Section 4.3,

  $$\Lambda_a{}^b \leftrightarrow e_a{}^\mu, \quad \Lambda^{-1}{}^b{}_c \leftrightarrow e^b{}_\nu.$$

  Equation (5.18) then becomes

  $$\Gamma_\mu{}^b{}_a = e^b{}_\nu \partial_\mu e_a{}^\nu + e_a{}^\nu \bar{\Gamma}_\mu{}^\rho{}_\nu e^b{}_\rho = -e_a{}^\nu \partial_\mu e^b{}_\nu + e_a{}^\nu \bar{\Gamma}_\mu{}^\rho{}_\nu e^b{}_\rho, \tag{5.19}$$

  where the second inequality comes from the fact that the contraction $e^b{}_\nu e_a{}^\nu = \delta^b_a$ has a vanishing derivative. This equation can be rewritten suggestively as

  $$\partial_\mu e^b{}_\rho - \bar{\Gamma}_\mu{}^\nu{}_\rho e^b{}_\nu + \Gamma_\mu{}^b{}_a e^a{}_\rho = 0, \tag{5.20}$$

  which is the condition for the vanishing of the covariant derivative of $e_a{}^\mu$, with appropriate connection components for each index.

5.5 The Christoffel/Levi-Civita connection

So far, the connection has been essentially arbitrary. We will now impose two conditions to restrict its form. To do so, we need two new tensors, torsion and nonmetricity.

Let f be an arbitrary scalar. The first covariant derivative, $\nabla_\nu f$, doesn't involve the connection—it's just the ordinary derivative—but the second covariant derivative, $\nabla_\mu \nabla_\nu f$, does. In a coordinate basis, it is simple to compute the commutator,

$$[\nabla_\mu, \nabla_\nu]f = (\nabla_\mu \nabla_\nu - \nabla_\nu \nabla_\mu)f = -T^\rho_{\mu\nu} \nabla_\rho f \quad \text{with} \quad T^\rho_{\mu\nu} = \Gamma^\rho_{\mu\nu} - \Gamma^\rho_{\nu\mu}. \tag{5.21}$$

The object $T^\rho_{\mu\nu}$ is called the torsion. Equation (5.21) is a tensor equation; although the connection is not a tensor, the torsion, a difference between two connections, is.

Now let $g_{\mu\nu}$ be a metric, and consider its covariant derivative. In a coordinate basis,

$$\nabla_\rho g_{\mu\nu} = \partial_\rho g_{\mu\nu} - g_{\sigma\nu} \Gamma^\sigma_{\rho\mu} - g_{\mu\sigma} \Gamma^\sigma_{\rho\nu} = N_{\rho\mu\nu}. \tag{5.22}$$

The object $N_{\rho\mu\nu}$ is called the nonmetricity. It, too, is a tensor.

A geometry in which the the torsion and nonmetricity are both set to zero is called Riemannian if the metric is positive definite, or pseudo-Riemannian otherwise. These conditions completely determine the connection. Vanishing torsion means that Γ, in a coordinate basis, is symmetric in its two lower indices. The equation for vanishing nonmetricity, or "metric compatibility," can then be solved algebraically; by taking sums of (5.22) with suitably permuted indices, one obtains

$$\Gamma^\rho_{\mu\nu} = \Gamma^\rho_{\nu\mu} = \frac{1}{2} g^{\rho\sigma} (\partial_\mu g_{\sigma\nu} + \partial_\nu g_{\sigma\mu} - \partial_\sigma g_{\mu\nu}). \tag{5.23}$$

This "torsion-free, metric-compatible" connection is sometimes called the Christoffel connection, after Elwin Bruno Christoffel, who first introduced it, or the Levi-Civita connection, after Tullio Levi-Civita, who further elaborated the idea. The order of indices in (5.23) is conventional: we don't normally raise or lower indices (since $\Gamma^\rho_{\mu\nu}$ is not a tensor), so the relative position of the upper and lower indices doesn't matter.

This connection is exactly the object (2.40) that appears in the geodesic equation. In fact, the geodesic equation can now be written in the simple form

$$u^\nu \nabla_\nu u^\mu = 0 \quad \text{with} \quad u^\mu = \frac{dx^\mu}{ds}. \tag{5.24}$$

This is the "autoparallel" condition, the statement that the tangent vector $u = u^\mu \partial_\mu$ remains constant—and therefore parallel to itself—along a geodesic. This gives a mathematical justification for the vanishing of nonmetricity: it guarantees that the geodesics are the same as the autoparallel curves. As promised in Section 2.4, this autoparallel description applies just as well for null geodesics, with no need to worry about any limiting process.

The form of the Levi-Civita connection allows a simplification of certain covariant derivatives. We start with a fundamental matrix identity, Jacobi's formula: if M is an invertible matrix and δ is any derivative, then

$$\delta(\ln \det M) = \text{Tr}(M^{-1} \delta M). \tag{5.25}$$

> **Box 5.2 Do geodesics really determine geometry?**
>
> In Chapter 1 it was argued that the principle of equivalence—the uniqueness of trajectories in a gravitational field—meant that gravity could be viewed as geometry. We can now ask more specifically how uniquely the geodesics determine the geometry. This question was largely answered by Ehlers, Pirani, and Schild [39], who showed
>
> 1. The null geodesics determine the topology and the conformal geometry, that is, the metric up to a local rescaling (a "Weyl transformation") $g_{\rho\sigma} \to e^\lambda g_{\rho\sigma}$, where λ is an arbitrary function.
> 2. The timelike geodesics determine a projective structure, that is, a connection up to a transformation $\Gamma^\rho_{\mu\nu} \to \Gamma^\rho_{\mu\nu} + \frac{1}{2}(\delta^\rho_\mu \alpha_\nu + \delta^\rho_\nu \alpha_\mu)$.
> 3. Compatibility—the fact that trajectories of massive objects lie within light cones but can "chase" light arbitrarily closely—implies that $\nabla_\mu g_{\rho\sigma} = \alpha_\mu g_{\rho\sigma}$ for some vector α.
> 4. The observation that the rate of a clock doesn't depend on its history (no "second clock effect") implies that $\alpha_\mu = -\partial_\mu \omega$ for some function ω.
> 5. The rescaled metric $\tilde{g}_{\rho\sigma} = e^\omega g_{\rho\sigma}$ then satisfies $\nabla_\mu \tilde{g}_{\rho\sigma} = 0$; that is, it is compatible with the connection. This metric reproduces both the null and the timelike geodesics, and is essentially unique.

In particular, if we take M to be the matrix $||g||$ of components of the metric, with determinant $\det ||g|| = g$, then

$$g^{-1}\delta g = \text{Tr}\, ||g||^{-1} \delta ||g|| = g^{\mu\nu} \delta g_{\mu\nu} \Rightarrow \delta\sqrt{|g|} = \frac{1}{2}\sqrt{|g|}\, g^{\mu\nu} \delta g_{\mu\nu}. \tag{5.26}$$

Now consider the Levi-Civita connection contracted on one index. From (5.23),

$$\Gamma^\rho_{\mu\rho} = \frac{1}{2} g^{\rho\sigma} \partial_\mu g_{\rho\sigma} = \frac{1}{\sqrt{|g|}} \partial_\mu \sqrt{|g|}. \tag{5.27}$$

This enormously simplifies the covariant divergence of a tangent vector:

$$\nabla_\rho v^\rho = \partial_\rho v^\rho + \Gamma^\rho_{\mu\rho} v^\mu = \frac{1}{\sqrt{|g|}} \partial_\mu \left(\sqrt{|g|}\, v^\mu\right). \tag{5.28}$$

This last equality allows us to integrate by parts even with integrands containing covariant derivatives. For example, consider an integral

$$\int_M S^{\mu\nu} \nabla_\mu \omega_\nu \sqrt{|g|}\, d^d x,$$

where we saw in Section 5.1 that the factor of $\sqrt{|g|}$ was needed for the expression to make mathematical sense. We can now write

$$\int_M S^{\mu\nu}\nabla_\mu \omega_\nu \sqrt{|g|}\, d^d x = \int_M \{\nabla_\mu(S^{\mu\nu}\omega_\nu) - (\nabla_\mu S^{\mu\nu})\omega_\nu\}\sqrt{|g|}\, d^d x$$

$$= -\int_M (\nabla_\mu S^{\mu\nu})\omega_\nu \sqrt{|g|}\, d^d x + \int_M \partial_\mu(\sqrt{|g|} S^{\mu\nu}\omega_\nu)\, d^d x$$

$$= -\int_M (\nabla_\mu S^{\mu\nu})\omega_\nu \sqrt{|g|}\, d^d x + \text{boundary terms}, \qquad (5.29)$$

where, just as in ordinary calculus, the integral of a total derivative gives only boundary terms. This manipulation is a version of Stokes' theorem, which is discussed in more mathematical detail in Section A.5. The generalization to other types of tensors is straightforward, and tells us that as long as the integrand is a scalar density, we can integrate by parts with covariant derivatives just as we would with ordinary derivatives.

5.6 Parallel transport

In the form (5.24), the geodesic equation is the statement that the tangent vector of a geodesic is "parallel transported," that is, carried along the geodesic in such a way that it remains parallel to itself. This notion can be generalized to arbitrary vectors transported along arbitrary curves. Given a parametrized curve $x^\mu(\sigma)$, a vector v^a is said to be parallel transported along the curve if

$$\frac{dx^\mu}{d\sigma}\nabla_\mu v^a = 0. \qquad (5.30)$$

Parallel transport has several important features:

- If two vectors v and w are parallel transported along the same curve, their inner product is preserved:

$$\frac{dx^\mu}{d\sigma}\nabla_\mu(g_{ab}v^a w^b) = g_{ab} v^a \frac{dx^\mu}{d\sigma}\nabla_\mu w^b + g_{ab} w^b \frac{dx^\mu}{d\sigma}\nabla_\mu v^a + v^a w^b \frac{dx^\mu}{d\sigma}\nabla_\mu g_{ab} = 0, \qquad (5.31)$$

where the first two terms vanish because v and w are parallel transported and the third vanishes because $\nabla_\mu g_{ab} = 0$.
- In particular, the length of a vector, $g_{ab}v^a v^b$, is preserved. This is a further reason to assume metric compatibility; if $\nabla_\mu g_{ab}$ were not zero, the length of a vector would depend on its history (the "second clock effect" mentioned in Box 5.2).
- If a vector v is parallel transported along a geodesic, its angle with the geodesic remains constant. Since the metric defines the inner product, we can define the angle between vectors u and v by the usual formula

$$\cos\theta = \frac{g_{ab} u^a v^b}{|u||v|} \quad \text{with } |u|^2 = g_{ab} u^a u^b,\ |v|^2 = g_{ab} v^a v^b. \qquad (5.32)$$

If we take u to be the tangent vector to a geodesic, then both u and v are parallel transported, v by assumption, u because of the geodesic equation. Hence $|u|$, $|v|$ and $g_{ab}u^a v^b$ are all constant, and so is the angle (5.32).

5.7 Curvature

We now have the machinery we need to describe curvature. To set the stage, let us begin with an apparent paradox.

Pick a point $p \in M$ and a unit vector v_0 at p. Using parallel transport, as defined in the preceding section, transport v_0 to each other point p' of M, creating a vector field $v_0(x)$. Since parallel transport preserves lengths, $v_0(x)$ will be a unit vector at each point. Now choose a new unit vector v_1 at p that is orthogonal to v_0, and repeat. Since parallel transport preserves inner products, $v_1(x)$ will remain orthogonal to $v_0(x)$ at each point. Repeat d times.

We now have an orthonormal frame at each point. But it's more than that: it's a Cartesian grid, a frame in which each vector v_a at the point p is parallel to the corresponding vector at every other point. A manifold that admits such a structure is flat—it's ordinary Euclidean or Minkowski space, complete with Cartesian coordinates. It seems that in defining parallel transport, we have accidentally limited ourselves to flat manifolds.

We haven't, of course. The root of the paradox can be traced back to the definition (5.30). Parallel transport is defined *along a particular curve*. If the transport of a vector from p to p' depends on the curve, the first step of the paradox fails: we don't really have a single-valued vector field $v_0(x)$. This gives us our first definition of curvature:

- **Curvature is the obstruction to existence of covariantly constant vectors.**

Equation (5.30) defines parallel transport along a curve with a given tangent vector. For the result to be independent of the curve, the relation must hold for all possible tangent vectors, so

$$\nabla_\mu v^a = \partial_\mu v^a + \Gamma_\mu{}^a{}_b v^b = 0. \tag{5.33}$$

A vector field satisfying (5.33) is said to be covariantly constant. From the theory of differential equations, the integrability condition for this equation—the necessary and sufficient condition for a local solution to exist—is that $(\partial_\mu \partial_\nu - \partial_\nu \partial_\mu) v^a = 0$, that is,

$$\partial_\mu \left(\Gamma_\nu{}^a{}_b v^b \right) - \partial_\nu \left(\Gamma_\mu{}^a{}_b v^b \right) = 0$$
$$= (\partial_\mu \Gamma_\nu{}^a{}_b) v^b + \Gamma_\nu{}^a{}_b \partial_\mu v^b - (\partial_\nu \Gamma_\mu{}^a{}_b) v^b - \Gamma_\mu{}^a{}_b \partial_\nu v^b$$
$$= (\partial_\mu \Gamma_\nu{}^a{}_b) v^b - \Gamma_\nu{}^a{}_b \Gamma_\mu{}^b{}_c v^c - (\partial_\nu \Gamma_\mu{}^a{}_b) v^b + \Gamma_\mu{}^a{}_b \Gamma_\nu{}^b{}_c v^c. \tag{5.34}$$

The condition for the existence of a covariantly constant vector v^a is thus

$$R_{\mu\nu}{}^a{}_b v^b = 0, \tag{5.35}$$

where

$$R_{\mu\nu}{}^a{}_b = \partial_\mu \Gamma_\nu{}^a{}_b - \partial_\nu \Gamma_\mu{}^a{}_b + \Gamma_\mu{}^a{}_c \Gamma_\nu{}^c{}_b - \Gamma_\nu{}^a{}_c \Gamma_\mu{}^c{}_b. \tag{5.36}$$

The object $R_{\mu\nu}{}^a{}_b$ is called the Riemann curvature tensor, after the mathematician Bernhard Riemann, or the curvature tensor for short. Equation (5.35) depends on the vector v^a, but for a full basis of vectors to be covariantly constant, as required for flatness, we would need the full curvature tensor to vanish, $R_{\mu\nu}{}^a{}_b = 0$.

A second definition of curvature is closely related:

- **Curvature measures the change of a vector transported around a curve.**

  If parallel transport depends on the choice of curve, then parallel transport around a closed curve will not necessarily bring a vector back to its initial value. Let $x^\mu(\sigma)$ be a closed curve, with $\sigma \in [0,1]$ and $x^\mu(0) = x^\mu(1)$. Choose a vector $v^a(0)$ at the starting point $\sigma = 0$ and parallel transport it along the curve. The final value $v^a(1)$ will have the form

  $$v^a(1) = H^a{}_b v^b(0), \tag{5.37}$$

  where the matrix $H^a{}_b$ is called the holonomy of the curve. The holonomy is calculated in Section A.6; for a curve of length $\ell \ll 1$,

  $$H^a{}_b = \delta^a_b + \int_\Sigma R_{\mu\nu}{}^a{}_b \, d\sigma^{\mu\nu} + \mathcal{O}(\ell^3), \tag{5.38}$$

  where the integral is over any surface enclosed by the curve.

A third definition of curvature comes from another measure of the failure of flatness:

- **Curvature measures the noncommutativity of covariant derivatives.**

  If a manifold is flat, we can find Cartesian coordinates, in which the components of the metric in a coordinate basis are constant. In such a coordinate system, the components of the Levi-Civita connection are zero, so $\nabla_\mu = \partial_\mu$. Thus for any vector v, the commutator of covariant derivatives is $[\nabla_\mu, \nabla_\nu] v^a = 0$. While this equality was derived in a particular coordinate system, it is a tensor equation, so it holds in any coordinate system. Conversely, if $[\nabla_\mu, \nabla_\nu] v^a \neq 0$ for even one vector, the manifold cannot be flat.

  A simple calculation from the definition (5.15) of the covariant derivative now gives

  $$[\nabla_\mu, \nabla_\nu] v^a = R_{\mu\nu}{}^a{}_b v^b. \tag{5.39}$$

  The curvature again measures the degree to which the manifold fails to be flat.

  Note that for a metric-compatible connection, we can pass the metric through the covariant derivatives in (5.39) to raise and lower indices. Thus, for instance,

  $$[\nabla_\mu, \nabla_\nu] \omega_a = R_{\mu\nu a}{}^b \omega_b, \tag{5.40}$$

  with similar equations for other types of tensors.

A final definition of curvature is slightly different, but ties in more closely with physics. In flat spacetime, two objects moving along parallel geodesics will remain a constant distance apart. In a curved spacetime describing a gravitational field, they will tend to converge.

- **Curvature is a measure of geodesic deviation.**

  To set up the problem, consider a sheet of geodesics, as shown in Fig. 5.1, where the proper time along each geodesic is s and the distinct geodesics are labeled by a parameter t.

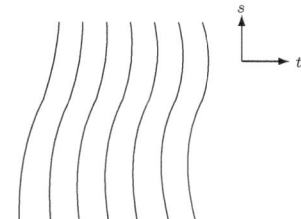

Fig. 5.1 A sheet of geodesics

Define the velocity and separation vectors

$$u = \frac{\partial}{\partial s} = \frac{\partial x^\mu}{\partial s}\partial_\mu, \quad X = \frac{\partial}{\partial t} = \frac{\partial x^\mu}{\partial t}\partial_\mu, \tag{5.41}$$

where u satisfies the geodesic equation. The relative velocity V and relative acceleration A of neighboring geodesics are then

$$V^\mu = u^\nu \nabla_\nu X^\mu, \quad A^\mu = u^\nu \nabla_\nu V^\mu = u^\nu \nabla_\nu (u^\rho \nabla_\rho X^\mu). \tag{5.42}$$

We will need the identity

$$u^\nu \nabla_\nu X^\mu - X^\nu \nabla_\nu u^\mu = u^\nu \partial_\nu X^\mu - X^\nu \partial_\nu u^\mu = \frac{\partial}{\partial s}\frac{\partial x^\mu}{\partial t} - \frac{\partial}{\partial t}\frac{\partial x^\mu}{\partial s} = 0, \tag{5.43}$$

where the first equality comes from direct calculation—for a torsion-free connection the terms containing Γ cancel—and the second comes from the definitions (5.41). We now compute the relative acceleration:

$$A^\mu = u^\nu \nabla_\nu (u^\rho \nabla_\rho X^\mu) = u^\nu \nabla_\nu (X^\rho \nabla_\rho u^\mu) = (u^\nu \nabla_\nu X^\rho)\nabla_\rho u^\mu + X^\rho u^\nu \nabla_\nu \nabla_\rho u^\mu$$

using (5.39) to commute covariant derivatives:

$$= (X^\nu \nabla_\nu u^\rho)\nabla_\rho u^\mu + X^\rho u^\nu (\nabla_\rho \nabla_\nu u^\mu + R_{\nu\rho}{}^\mu{}_\sigma u^\sigma)$$
$$= (X^\nu \nabla_\nu u^\rho)\nabla_\rho u^\mu + X^\rho [\nabla_\rho (u^\nu \nabla_\nu u^\mu) - (\nabla_\rho u^\nu)(\nabla_\nu u^\mu)] + R_{\nu\rho}{}^\mu{}_\sigma u^\sigma X^\rho u^\nu$$

using the geodesic equation $u^\nu \nabla_\nu u^\mu = 0$, and relabel dummy indices:

$$= R_{\nu\rho}{}^\mu{}_\sigma u^\sigma X^\rho u^\nu. \tag{5.44}$$

The relative acceleration of neighboring geodesics—the geodesic deviation—is thus given by the curvature tensor. We shall see in the next chapter that this fact has a close connection to the Einstein field equations.

5.8 Properties of the curvature tensor

In d dimensions, the curvature tensor has d^4 components. Fortunately for us, algebraic identities greatly reduce the number of independent components. For the curvature associated with the Levi-Civita connection,

$$R_{\mu\nu}{}^a{}_b = -R_{\nu\mu}{}^a{}_b,\tag{5.45}$$
$$R_{\mu\nu ab} = -R_{\mu\nu ba},\tag{5.46}$$
$$R_{\mu\nu\rho\sigma} + R_{\nu\rho\mu\sigma} + R_{\rho\mu\nu\sigma} = 0 \quad \text{in a coordinate basis},\tag{5.47}$$
$$R_{\mu\nu\rho\sigma} = R_{\rho\sigma\mu\nu} \quad \text{in a coordinate basis}.\tag{5.48}$$

The effect is to reduce the number of independent components to $\frac{1}{12}d^2(d^2-1)$.

The first of these identities follows trivially from (5.39). The second comes from metric compatibility:

$$[\nabla_\mu, \nabla_\nu]g_{ab} = R_{\mu\nu}{}^c{}_a g_{cb} + R_{\mu\nu}{}^c{}_b g_{ac} = 0.$$

For the third, we need the Jacobi identity, a very general algebraic identity for commutators,

$$[[A, B], C] + [[B, C], A] + [[C, A], B] = 0,\tag{5.49}$$

which can be verified by simply expanding the commutators and confirming that all of the terms cancel in pairs. Now let A, B, and C be covariant derivatives acting on a scalar f. From the torsion-free condition, $[\nabla_\mu, \nabla_\nu]f = 0$, so

$$[[\nabla_\mu, \nabla_\nu], \nabla_\rho]f = [\nabla_\mu, \nabla_\nu](\nabla_\rho f) = R_{\mu\nu\rho}{}^\sigma \nabla_\sigma f,$$

which, inserted into the Jacobi identity, gives (5.47). The final identity (5.48) is an algebraic consequence of the other three.

The Jacobi identity also implies a differential identity for the curvature tensor. Again let A, B, and C be covariant derivatives, but now acting on a vector v. One then finds

$$\nabla_\mu R_{\nu\rho}{}^a{}_b + \nabla_\nu R_{\rho\mu}{}^a{}_b + \nabla_\rho R_{\mu\nu}{}^a{}_b = 0.\tag{5.50}$$

This relation is known as the Bianchi identity (technically the "second Bianchi identity"—the first is (5.47)—but that terminology is rarely used).

We can now make several new tensors from $R_{\mu\nu}{}^a{}_b$ by contracting indices. Because of the symmetries (5.45)–(5.48), there is only one independent contraction to a rank two tensor,

$$R_{\mu\nu} = R_{\lambda\mu}{}^\lambda{}_\nu.\tag{5.51}$$

This is the Ricci tensor, named after the mathematician Gregorio Ricci-Curbastro. From (5.48), $R_{\mu\nu}$ is symmetric. As described in Box 6.1, the Ricci tensor characterizes the deviation of the volume element $\sqrt{-g}\,d^4x$ from that of flat spacetime. Physically, this property can be encoded in the rate of expansion of a "pencil" of geodesics, whose cross-sectional area is described by a relation known as the Raychaudhuri equation, a generalization of the geodesic deviation equation (5.44). Roughly speaking, positive eigenvalues of the Ricci tensor cause convergence of geodesics and a decrease in volume, while negative eigenvalues cause divergence.

We can further contract the Ricci tensor to form a scalar,

$$R = R^\mu{}_\mu = g^{\mu\nu}R_{\mu\nu},\tag{5.52}$$

which is known to physicists as the curvature scalar and to mathematicians as the scalar curvature. The combination

Box 5.3 Conventions: part IV

Conventions for the curvature tensor and its contractions are unfortunately not standard. The definition (5.36) is the most common one, but some authors define the curvature tensor with the opposite sign. The Ricci tensor (5.51) was defined here as a contraction of the first and third indices of the curvature tensor; other authors contract the first and fourth, leading to the opposite sign for the Ricci tensor and the curvature scalar. In a coordinate basis, some authors switch the position of the first and second pair of indices of $R_{\mu\nu\rho\sigma}$. The identity (5.48) makes this unimportant for a torsion-free, metric-compatible connection, but in more general cases it can matter.

The Levi-Civita connection, curvature tensor, and Ricci tensor are independent of the sign of the metric. The curvature scalar switches sign between the East Coast and West Coast conventions, however (see Box 2.2). Again, one must pay close attention to ensure consistency.

$$G_{\mu\nu} = R_{\mu\nu} - \frac{1}{2} g_{\mu\nu} R \tag{5.53}$$

is the Einstein tensor. It is "covariantly conserved"—it follows from an appropriate contraction of the Bianchi identity (5.50) that

$$\nabla_\mu G^{\mu\nu} = 0. \tag{5.54}$$

The Ricci tensor is the "trace" part of the curvature. In four dimensions, it carries half the total information—the full curvature tensor has twenty independent components, while the Ricci tensor has ten. The remaining components constitute the Weyl tensor $C_{\mu\nu\rho\sigma}$ (also sometimes denoted $W_{\mu\nu\rho\sigma}$), named for the physicist and mathematician Hermann Weyl. In d dimensions,

$$C_{\mu\nu\rho\sigma} = R_{\mu\nu\rho\sigma} + \frac{2}{d-2} \left(g_{\mu[\rho} R_{\sigma]\nu} - g_{\nu[\rho} R_{\sigma]\mu} \right) - \frac{2}{(d-1)(d-2)} R g_{\mu[\rho} g_{\sigma]\nu}. \tag{5.55}$$

The Weyl tensor has the same symmetries (5.45–5.48) as the curvature tensor, and the coefficients are chosen so that it is traceless, that is, the contraction of any pair of indices gives zero. It is also invariant under local rescalings: under a transformation of the metric of the form

$$g_{\mu\nu} \to e^{2\omega} g_{\mu\nu}, \tag{5.56}$$

the components $C_{\mu\nu}{}^\rho{}_\sigma$ (the index positions matter) are unchanged.

A rescaling of the form (5.56) is called a Weyl transformation or, loosely, a conformal transformation. While such a rescaling changes the geometry, it preserves angles and—important for the discussion of Penrose diagrams coming in Section 10.6—it leaves null geodesics unchanged.

5.9 The Cartan structure equations

To finish, let us translate some of these results into the language of orthonormal bases of Section 4.7. We start with the Levi-Civita connection. In an orthonormal basis, this connection is traditionally written as $\omega_\mu{}^a{}_b$, and is called the "spin connection," since it was first introduced in physics to handle covariant derivatives of spinors.

The metric compatibility condition is now simple: the components η_{ab} of the metric are constant in an orthonormal basis, so

$$\nabla_\mu g_{ab} = \partial_\mu \eta_{ab} - \eta_{cb}\,\omega_\mu{}^c{}_a - \eta_{ac}\,\omega_\mu{}^c{}_b = 0 \;\Rightarrow\; \omega_{\mu ab} = -\omega_{\mu ba}\,. \tag{5.57}$$

For the torsion-free condition, we can use (5.20) to transform from a coordinate basis to an orthonormal basis. The vanishing of torsion in the coordinate basis then implies

$$(\partial_\mu e^a{}_\nu + \omega_\mu{}^a{}_b\,e^b{}_\nu) - (\partial_\nu e^a{}_\mu + \omega_\nu{}^a{}_b\,e^b{}_\mu) = 0\,. \tag{5.58}$$

In principle, we can find a full spin connection, by inserting the Christoffel connection (5.23) into (5.19). For our purposes here, though, (5.58) will suffice.

We can now elegantly reexpress these relations in the language of forms. Define three differential forms:

- the basis one-form (or "coframe")

$$e^a = e^a{}_\mu\,dx^\mu\,, \tag{5.59}$$

- the connection one-form

$$\omega^a{}_b = \omega_\mu{}^a{}_b\,dx^\mu\,, \tag{5.60}$$

- the curvature two-form

$$\mathcal{R}^a{}_b = \frac{1}{2} R_{\mu\nu}{}^a{}_b\,dx^\mu \wedge dx^\nu\,. \tag{5.61}$$

In the language of Cartan calculus, the condition of metric compatibility is then

$$\omega_{ab} = -\omega_{ba}\,, \tag{5.62}$$

the torsion-free condition (5.58) is

$$de^a + \omega^a{}_b \wedge e^b = 0 \qquad \text{(Cartan's first structure equation)}, \tag{5.63}$$

and the curvature tensor (5.36) is

$$\mathcal{R}^a{}_b = d\omega^a{}_b + \omega^a{}_c \wedge \omega^c{}_b \qquad \text{(Cartan's second structure equation)}. \tag{5.64}$$

Using the identity (A.10), we can also find a simple expression for the curvature scalar:

$$\tilde{\epsilon}_{abcd}\,e^a \wedge e^b \wedge \mathcal{R}^{cd} = 2\sqrt{|g|}\,R\,d^4x\,. \tag{5.65}$$

The formalism is not merely elegant; it is also a very powerful calculational tool. Consider, for example, the two-sphere with the metric (2.16). We can read off an orthonormal basis from the line element,

$$e^1 = r_0 d\theta, \quad e^2 = r_0 \sin\theta d\varphi. \tag{5.66}$$

Because of antisymmetry, the connection one-form in two dimensions has only one independent component $\omega_{12} = \omega$. Since our metric has Riemannian signature, its components in an orthonormal basis are $\eta_{ab} = \delta_{ab}$, and

$$\omega^1{}_2 = \eta^{11}\omega_{12} = \omega, \quad \omega^2{}_1 = \eta^{22}\omega_{21} = -\eta^{22}\omega_{12} = -\omega.$$

The first structure equation is thus

$$de^1 + \omega^1{}_2 \wedge e^2 = r_0 \sin\theta\, \omega \wedge d\varphi = 0,$$
$$de^2 + \omega^2{}_1 \wedge e^1 = r_0 \cos\theta\, d\theta \wedge d\varphi - r_0 \omega \wedge d\theta = 0, \tag{5.67}$$

since $d^2 = 0$. From this, with a little practice, we can read off

$$\omega = -\cos\theta\, d\varphi. \tag{5.68}$$

By antisymmetry, the curvature two-form also has only one independent component,

$$\mathcal{R}^1{}_2 = d\omega^1{}_2 + \omega^1{}_c \wedge \omega^c{}_2 = d\omega = \sin\theta\, d\theta \wedge d\varphi. \tag{5.69}$$

From the definition (5.61), this means

$$R_{\theta\varphi}{}^1{}_2 = \sin\theta. \tag{5.70}$$

To transform to a coordinate basis, we simply write

$$R_{\theta\varphi 12}\, e^1 \otimes e^2 = R_{\theta\varphi\theta\varphi}\, e^\theta \otimes e^\varphi \Rightarrow R_{\theta\varphi\theta\varphi} = r_0^2 \sin^2\theta = \frac{1}{r_0^2} g_{\theta\theta}g_{\varphi\varphi}. \tag{5.71}$$

In higher dimensions, calculations are more complicated, of course, and in Lorentzian signature one must keep track of signs. But the Cartan structure equations are still often the simplest way to compute curvature, as we shall see in Section 10.3.

Further reading

The references at the end of Chapter 4— in particular, [11], [33], [35], and [36]—are also useful here. Some more technical details have again been relegated to Appendix A. The Cartan structure equations first appeared in [40]. As described in [33], the concept of curvature can be extended to the more general setting of a fiber bundle, where it can be used to describe electromagnetic and Yang-Mills field strengths. For a nice physicists' overview of this broader setting, see [41]. The Raychaudhuri equation is described in detail in Poisson's book, *A Relativist's Toolkit* [42].

6
The Einstein field equations

We now have the mathematical equipment to write down the Einstein field equations, the way "matter tells spacetime how to curve." The most elegant derivation starts from an action principle, which is uniquely determined by a simple set of assumptions. First, though, it's helpful to get some physical intuition from Newtonian gravity.

The two preceding chapters developed differential geometry for arbitrary spacetime dimensions. To keep notation simple, the remainder of this book will assume $d = 4$ unless otherwise noted. Generalizations to other dimensions are almost always straightforward, requiring only some changes in constants.

6.1 Gravity and geodesic deviation

Newtonian gravity is a theory of acceleration. More precisely, it is a theory of *relative* acceleration. An astronaut in an orbiting spacecraft is weightless; gravity can only be seen by comparing the motion of the spacecraft to the Earth, or perhaps by looking for tiny differences at opposite ends of the spacecraft. So a natural starting place is the geometrical description of relative acceleration, the geodesic deviation equation of Section 5.7.

In Newtonian gravity, the "absolute" acceleration of a body in an inertial frame is

$$a^i = \eta^{ik} \partial_k \Phi, \tag{6.1}$$

where Φ is the gravitational potential and $i, j, k, \ldots$ are spatial indices. (Recall that with our "West Coast" sign convention, $\eta^{ij} = -\delta^{ij}$.) The relative acceleration of two neighboring bodies separated by a distance δx^k is therefore

$$A^i_{\text{Newton}} = \eta^{ik} \delta x^j \partial_j \partial_k \Phi. \tag{6.2}$$

This is to be compared to equation (5.44) for geodesic deviation. For an object moving much more slowly than light, the velocity u^μ has components $u^0 = 1 + \mathcal{O}(v^2/c^2)$ and $u^i = \mathcal{O}(v/c)$, so to lowest order in v/c

$$A^i \approx R_{0j}{}^i{}_0 X^j. \tag{6.3}$$

To reproduce Newtonian gravity geometrically, we thus need

$$R_{0j}{}^i{}_0 \sim \eta^{ik} \partial_j \partial_k \Phi. \tag{6.4}$$

In particular, contracting the indices and using the Poisson equation for the Newtonian potential, we expect

$$R_{00} \sim \nabla^2 \Phi = 4\pi G \rho, \tag{6.5}$$

where ρ is the mass density.

This is certainly not the final form of the field equations—at best, it's one component of what must be a tensor equation. But it suggests the general form: some contracted version of the curvature tensor should equal some object representing mass or energy density. Einstein's original derivation came from searching for an equation of this form. In the next section, we will take the more elegant path proposed by David Hilbert, but (6.5) will remain as a vital check that the results make physical sense.

6.2 The Einstein-Hilbert action

The principle of least action is ubiquitous in physics. The equations governing fundamental physical interactions almost always come from an action principle, in which field equations are found by extremizing a functional $I[\psi]$ of fields ψ. This procedure might have its origin in the quantum mechanical path integral, where each "path" has a phase $I[\psi]/\hbar$ and the extrema determine a stationary phase approximation. But whatever the ultimate explanation, it is natural to look for an action principle for spacetime geometry.

A local action is an integral over spacetime of some function of the dynamical variables and their derivatives. From the last two chapters, we know that our fundamental variables are the metric g and the connection Γ. Let us impose a few simple conditions:

1. There should be no "background structure"—the gravitational part of the action should depend only on the geometry. In particular, there should be no preferred coordinate system or reference frame.
2. The geometry should be pseudo-Riemannian; that is, Γ should be the Levi-Civita connection. As described in Box 5.2, the trajectories of particles and light—timelike and null geodesics—determine a unique metric and a compatible connection; the assumption is that this is the geometry relevant to the action.
3. To ensure that the equations of motion are second order, the action should involve no terms containing more than two derivatives of the metric and should be linear in second derivatives (see the discussion of Ostrogradsky's theorem in Box 1.1).

By condition 1, the gravitational action must be of the form

$$I = \int_M f \sqrt{|g|}\, d^4x$$

where f is a scalar function of the metric and connection. By condition 3, no term in f should have more than two derivatives. The only possibilities are then:

- **No derivatives:** There is no nonconstant scalar that can be constructed algebraically from the metric, but f can contain a constant term.
- **One derivative:** No scalar can be built from the metric and only single derivatives. Suppose χ were such a scalar. Pick a point p and a freely falling frame at p (see Box 6.1). In that frame, $\partial_\rho g_{\mu\nu} = 0$ at p, so $\chi(p)$ must be a constant, with no metric dependence. But this is a coordinate-independent statement, so it must hold in any frame. Since the point p was arbitrary, χ must be independent of the metric everywhere.
- **Two derivatives:** We have seen one scalar linear in second derivatives of the metric, the curvature scalar. It can be shown that this is the only such object.

Box 6.1 Freely falling frames

It is often convenient to transform to a freely falling reference frame, in which the effects of gravity at some point p disappear. Mathematically, this means choosing a special coordinate system, Riemann normal coordinates. In these coordinates, which always exist for a smooth enough metric, $g_{\mu\nu}(p) = \eta_{\mu\nu}$, $\partial_\rho g_{\mu\nu}(p) = 0$, and

$$g_{\mu\nu}(x) = \eta_{\mu\nu} - \frac{1}{3} R_{\mu\rho\nu\sigma}(x^\rho - x^\rho(p))(x^\sigma - x^\sigma(p)) + \ldots$$

near p. Equations involving the metric become much simpler in these coordinates, permitting a mathematical trick: if we can show that an equation holds in Riemann normal coordinates, and we can rewrite it as a relation among tensors, we can conclude that the tensorial form holds in any coordinate system.

As noted in Section 5.8, these coordinates also make it easier to understand the geometric significance of the Ricci tensor. It is easy to check that near p,

$$\sqrt{-g} = \sqrt{-\eta}\left(1 - \frac{1}{6} R_{\rho\sigma}(x^\rho - x^\rho(p))(x^\sigma - x^\sigma(p)) + \ldots\right),$$

so $R_{\rho\sigma}$ measures the deviation of the volume element from that of flat spacetime.

The general action that meets our criteria is thus

$$I = -\frac{1}{2\kappa^2} \int_M (R + 2\Lambda) \sqrt{|g|}\, d^4x + I_{\text{matter}}[g, \psi], \quad (6.6)$$

where κ and Λ are constants—Λ will turn out to be the cosmological constant—and ψ represents generic matter fields. This is the Einstein-Hilbert action.

Our next task is to extremize this action to obtain field equations. The matter term has been left general, so we can't say much about it, but we can define a symmetric tensor T_{ab} by the condition

$$\delta I_{\text{matter}}[g, \psi] = \frac{1}{2} \int_M T_{ab} \sqrt{|g|}\, \delta g^{ab}\, d^4x \quad (6.7)$$

under a variation of the metric. The tensor T_{ab} is called the stress-energy tensor (or energy-momentum tensor, or stress tensor, or stress-energy-momentum tensor). The next chapter will be devoted to understanding its properties.

For the gravitational term in the action, we have

$$\delta \int_M \left[(R + 2\Lambda)\sqrt{|g|}\right] d^4x$$
$$= \int_M \left[(R + 2\Lambda)\delta\sqrt{|g|} + \left((\delta g^{ab})R_{ab} + g^{ab}(\delta R_{ab})\right)\sqrt{|g|}\right] d^4x. \quad (6.8)$$

> **Box 6.2 Conventions: part V**
>
> The Einstein-Hilbert action (6.6) is one of the places in which sign conventions can be quite confusing. In the East Coast, "mostly plusses" convention common in cosmology, the sign of the curvature scalar changes, and the action becomes
>
> $$+\frac{1}{2\kappa^2}\int_M (R-2\Lambda)\sqrt{|g|}\, d^4x + I_{\text{matter}}[g,\psi].$$
>
> The sign of the stress-energy tensor in (6.7) also changes, and the net result is that Λ appears in the field equations (6.14) with the opposite sign. The physical meaning is still the same, though: positive Λ corresponds to de Sitter space (see Box 11.2), negative Λ to anti-de Sitter space.

We need the variations of $\sqrt{|g|}$ and of R_{ab}. For the first, we use Jacobi's formula (5.25):

$$\delta\sqrt{|g|} = \frac{1}{2}\sqrt{|g|}\, g^{ab}\delta g_{ab} = -\frac{1}{2}\sqrt{|g|}\, g_{ab}\delta g^{ab}. \quad (6.9)$$

(See (4.28) to understand the last sign change.) For the second, it is useful to work in a coordinate basis, in which

$$R_{\mu\nu} = \partial_\rho \Gamma^\rho_{\mu\nu} - \partial_\mu \Gamma^\rho_{\nu\rho} + \Gamma^\rho_{\mu\nu}\Gamma^\sigma_{\rho\sigma} - \Gamma^\rho_{\mu\sigma}\Gamma^\sigma_{\nu\rho}. \quad (6.10)$$

It may then be checked that

$$\delta R_{\mu\nu} = \partial_\rho \delta\Gamma^\rho_{\mu\nu} - \partial_\mu \delta\Gamma^\rho_{\nu\rho} + \delta\Gamma^\rho_{\mu\nu}\Gamma^\sigma_{\rho\sigma} + \Gamma^\rho_{\mu\nu}\delta\Gamma^\sigma_{\rho\sigma} - \delta\Gamma^\rho_{\mu\sigma}\Gamma^\sigma_{\nu\rho} - \Gamma^\rho_{\mu\sigma}\delta\Gamma^\sigma_{\nu\rho}$$
$$= \nabla_\rho \delta\Gamma^\rho_{\mu\nu} - \nabla_\mu \delta\Gamma^\rho_{\nu\rho}. \quad (6.11)$$

The $\partial\delta\Gamma$ and $\Gamma\delta\Gamma$ terms in the variation combine in just the right way to form covariant derivatives. This is not an accident. The variation of the connection is a difference, $\delta\Gamma = \Gamma[g+\delta g] - \Gamma[g]$, which we know from Section 5.4 is a tensor. Thus (6.11) is a tensor equation, and it is enough to verify it in a freely falling frame at a point p, as in Box 6.1. But in such a frame, $\Gamma(p) = 0$, and the equality is obvious.

Now, using the identity (5.28) for the covariant divergence, we have

$$\sqrt{|g|}\, g^{\mu\nu}\delta R_{\mu\nu} = \sqrt{|g|}\, \nabla_\rho v^\rho = \partial_\rho(\sqrt{|g|}\, v^\rho) \quad \text{with } v^\rho = g^{\mu\nu}\delta\Gamma^\rho_{\mu\nu} - g^{\rho\sigma}\delta\Gamma^\tau_{\sigma\tau}. \quad (6.12)$$

This is a total derivative, and will contribute only a boundary term to the variation. Such boundary terms are important for defining global quantities such as mass, and will be considered in Chapter 12. They are irrelevant for the "bulk" field equations, though; we can simply consider variations $\delta g^{\mu\nu}$ that go to zero at the boundaries.

Combining (6.8), (6.9), and (6.12), we thus find a total variation

$$\delta I_{\text{grav}} = -\frac{1}{2\kappa^2}\int_M \left[-\frac{1}{2}(R+2\Lambda)g_{ab} + R_{ab}\right]\sqrt{|g|}\,\delta g^{ab}\, d^4x. \quad (6.13)$$

56 The Einstein field equations

Adding the matter term (6.7) and setting the total variation to zero, we finally obtain

$$R_{ab} - \frac{1}{2}g_{ab}(R + 2\Lambda) = G_{ab} - \Lambda g_{ab} = \kappa^2 T_{ab}, \qquad (6.14)$$

where G_{ab} is the Einstein tensor (5.53). These are the Einstein field equations.

We can compare these equations to the earlier approximation (6.5). Apart from the cosmological constant Λ, the two are clearly of the same general form. To make the similarity even more evident, we can first contract the indices of (6.14), obtaining

$$-R - 4\Lambda = \kappa^2 T, \qquad (6.15)$$

where $T = g^{ab}T_{ab}$. Substituting back into (6.14), we find an equivalent "trace reversed" form of the Einstein field equations,

$$R_{ab} = \kappa^2 \left(T_{ab} - \frac{1}{2}g_{ab}T\right) - \Lambda g_{ab}. \qquad (6.16)$$

We will see in the next chapter that in the Newtonian approximation, $T_{00} \approx T \approx \rho$. So the time-time component of (6.16) will match the result of geodesic deviation (6.5) provided $\Lambda \sim 0$ and

$$\kappa^2 = 8\pi G. \qquad (6.17)$$

Note that we are using units in which $c = 1$. Many authors further choose units in which $G = 1$ ("natural" or "geometrized" units), and set $\kappa^2 = 8\pi$.

6.3 Conservation laws

We saw in Section 5.8 that the Einstein tensor satisfies the identity $\nabla_a G^{ab} = 0$. Thus when the field equations (6.14) are obeyed, the stress-energy tensor satisfies

$$\nabla_a T^{ab} = 0. \qquad (6.18)$$

This, as we will see in the next chapter, is a conservation law.

But (6.18) holds more widely: it is a direct consequence of the invariance of the matter action under coordinate transformations. Consider an infinitesimal transformation $x^\mu \to x^\mu + \xi^\mu$. As we worked out in Section 4.6, the metric changes under such a shift by

$$\delta_\xi g_{\mu\nu} = \partial_\mu \xi_\nu + \partial_\nu \xi_\mu - 2\xi_\rho \Gamma^\rho_{\mu\nu} = \nabla_\mu \xi_\nu + \nabla_\nu \xi_\mu. \qquad (6.19)$$

By (4.28), the inverse metric shifts with the opposite sign. The matter fields will also vary by some amount $\delta_\xi \psi$; we do not need the exact form.

The variation of the matter action under such a coordinate shift is then

$$\begin{aligned}\delta_\xi I_{\text{matter}} &= \frac{1}{2}\int_M T_{\mu\nu}\sqrt{|g|}\,\delta g^{\mu\nu}\,d^4x + \int_M \frac{\delta I_{\text{matter}}}{\delta \psi}\delta_\xi \psi\, d^4x \\ &= -\frac{1}{2}\int_M T_{\mu\nu}\sqrt{|g|}\,(\nabla^\mu \xi^\nu + \nabla^\nu \xi^\mu)\, d^4x + \int_M \frac{\delta I_{\text{matter}}}{\delta \psi}\delta_\xi \psi\, d^4x \\ &= \int_M \xi^\nu \nabla^\mu T_{\mu\nu}\sqrt{|g|}\,d^4x + \int_M \frac{\delta I_{\text{matter}}}{\delta \psi}\delta_\xi \psi\, d^4x, \qquad (6.20)\end{aligned}$$

where the last equality comes from integration by parts, as described in Section 5.5.

Box 6.3 Noether's theorem

The derivation (6.20) of the conservation law is a special case of a beautiful theorem of Emmy Noether, which states that any differentiable symmetry of an action is associated with a conservation law [43]. A sketch of the proof is as follows:

Start with an action $I = \int \mathcal{L}[\psi]\, d^d x$ with Lagrangian density $\mathcal{L}$. A variation $\hat{\delta}\psi$ of the fields will be a symmetry provided there is some functional $\Lambda[\psi]$ for which

$$\hat{\delta}\mathcal{L} = \partial_\mu \Lambda^\mu[\hat{\delta}\psi], \qquad (6.21)$$

since this reduces $\hat{\delta}I$ to a boundary integral. But it is also true that

$$\hat{\delta}\mathcal{L} = \frac{\partial \mathcal{L}}{\partial \psi}\hat{\delta}\psi + \frac{\partial \mathcal{L}}{\partial \partial_\mu \psi}\partial_\mu \hat{\delta}\psi = \partial_\mu\left(\frac{\partial \mathcal{L}}{\partial \partial_\mu \psi}\hat{\delta}\psi\right) + \left[\frac{\partial \mathcal{L}}{\partial \psi} - \partial_\mu\left(\frac{\partial \mathcal{L}}{\partial \partial_\mu \psi}\right)\right]\hat{\delta}\psi. \quad (6.22)$$

The last term is just the combination appearing in the Euler-Lagrange equations. So when the equations of motion are satisfied, (6.21) and (6.22) combine to give a continuity equation (see Section 7.5)

$$\partial_\mu\left(\frac{\partial \mathcal{L}}{\partial \partial_\mu \psi}\hat{\delta}\psi - \Lambda^\mu[\hat{\delta}\psi]\right) = 0 \quad \text{on shell.} \qquad (6.23)$$

But δ_ξ is just a shift of coordinates, so if the action is invariant, the variation (6.20) must be identically zero. Furthermore, the matter equations of motion are

$$\frac{\delta I_{\text{matter}}}{\delta \psi} = 0. \qquad (6.24)$$

So "on shell"—when the matter obeys its equations of motion—we must have

$$\int_M \xi^\nu \nabla^\mu T_{\mu\nu} \sqrt{|g|}\, d^4 x = 0$$

for every vector field ξ. This is exactly the conservation law (6.18).

6.4 Generalizing the action

As we have seen, the three conditions at the beginning of Section 6.2 enormously restrict the gravitational action. It is interesting to ask whether these conditions can be somehow loosened. A simple possibility, of course, is to introduce new nongeometrical fields—massless scalars, for instance—that mimic or distort the effects of geometry. Such models lie outside the scope of this book, although they will be briefly mentioned in the last chapter.

A more geometrical possibility is to drop the requirements of metric compatibility and vanishing torsion. The simplest approach, the Palatini variational principle, keeps the action (6.6), but treats the metric and connection as independent variables. This leads to a connection with nonzero nonmetricity and torsion, both determined by a new vector U_μ. As long as the stress-energy tensor is symmetric, though, U decouples and can be set to zero, and the field equations for the metric remain the same. A slightly different version in the spin connection formalism of Section 5.9 yields a torsion that couples to spinor fields, but does not propagate. Torsion and nonmetricity are both tensors, though, and once they are allowed, an enormous number of new terms can be added to the action. So far, no observational evidence has been found for the physical relevance of such terms, but investigations are continuing.

Another possibility is to allow more derivatives in the action. In greater than four spacetime dimensions, a more general action, the Lovelock action, incorporates a special set of higher order curvature polynomials whose variation still yields second order field equations. In four dimensions, higher derivative terms in the action necessarily lead to higher order field equations, but there are special combinations that can avoid the danger of unbounded energy discussed in Box 1.1. Again, there is no compelling observational evidence for such higher derivative terms, although a version due to Starobinsky led to one of the first models of inflationary cosmology.

Further reading

The final version of Einstein's field equations was published in [44]; a review paper he wrote a year later remains a very readable introduction [45]. Hilbert's action principle appeared in [46]. The proof that R is the only scalar built from the metric and its first and second derivatives that is linear in second derivatives goes back to Vermeil [47]. The extension to the most general action, in any dimension, that yields second order field equations is due to Lovelock [48].

Chapter 17 of [11] gives six different derivations of the field equations. Two more can be found in [49] (from quantum field theory) and [50] (from thermodynamics). Section 8.5 gives a derivation based on "bootstrapping" from flat spacetime.

While there is no doubt that it was Einstein's work that ultimately led to the field equations (6.14), there has been a dispute among historians of science as to whether Einstein or Hilbert derived the final form of the equations first. This now seems to be resolved in favor of Einstein [51].

Noether's theorem has application far beyond general relativity. For its use in quantum field theory, see, for example, Sections 7.3–7.4 of Weinberg's textbook, *Quantum Theory of Fields* [52].

The Palatini variational principle was originally published in [53]. It is described for metric variables in [54] and for tetrad variables in [55]. Theories with nonvanishing torsion are reviewed in [56]. The Lovelock action appeared in [48]. Starobinsky's early version of higher derivative actions, which remains interesting in cosmology, is found in [57].

7
The stress-energy tensor

The left-hand side of the Einstein field equations describes spacetime geometry. We now turn to the right-hand side, the stress-energy tensor, which describes the matter that acts as the "source" of gravity.

In eqn (6.7), we defined the stress-energy tensor as a functional derivative with respect to the metric. An obvious question is why such an object should have anything to do with energy. A partial answer comes from Section 6.3: the stress-energy tensor is the conserved quantity associated with shifts of coordinates. In particular, the components $T^\mu{}_0$ form a sort of conserved current associated with time translations. We know from both classical and quantum mechanics that the conserved generator of time translations is the Hamiltonian, which is also the energy. In the rest of this chapter we will try to understand this connection better.

7.1 Energy as a rank two tensor

Why should energy be part of a rank two tensor? To understand this, it's useful to start with the simpler case of electromagnetism. The source of the electric field is charge, Q, but the object that actually appears in Maxwell's equations is charge density, $\rho_{\text{charge}} = Q/V$, where V is volume. Under a Lorentz boost, Q is invariant, but V is Lorentz contracted in the direction of the boost, so

$$\rho_{\text{charge}} \to \gamma \rho_{\text{charge}},$$

where, as usual, $\gamma = (1 - v^2/c^2)^{-1/2}$. It is a basic feature of Lorentz symmetry that an object that transforms this way is the time component of a four-vector. So from special relativity alone, we know that charge density must be part of a vector J^μ.

To extend this argument to gravity, we must first decide on the equivalent of charge. There are two candidates for a relativistic version of gravitational mass, rest mass and energy. As a point of experimental fact, though, we know that all energy—nuclear binding energy, electrostatic energy, even the kinetic energy of electrons in atoms—contributes an amount E/c^2 to gravitational mass [2]. Thus the analog of charge density is $\rho_{\text{energy}} = E/V$. Now, under a Lorentz boost, the volume again contracts, but the energy also picks up a factor of γ, so

$$\rho_{\text{energy}} \to \gamma^2 \rho_{\text{energy}}.$$

An object that transforms in this way is the time-time component of a rank two tensor. So, again from special relativity alone, we know that energy density must be a part of a tensor $T^{\mu\nu}$.

General Relativity: A Concise Introduction. Steven Carlip © Steven Carlip 2019.
Published in 2019 by Oxford University Press. DOI: 10.1093/oso/9780198822158.001.0001

7.2 The stress-energy tensor for a point particle

As a first specific example of a stress-energy tensor, consider a point particle. The action I_{matter} is the quantity we extremize to get the equation of motion, the geodesic equation. Hence from Chapter 2, the action must be proportional to $\int ds$. In fact,

$$I_{\text{part}} = -m \int ds = -m \int \left(g_{\mu\nu} \frac{dx^\mu}{d\sigma} \frac{dx^\mu}{d\sigma} \right)^{1/2} d\sigma. \tag{7.1}$$

The coefficient is determined by the flat spacetime limit:

$$-mds = -m\sqrt{1-v^2}\, dt \approx \left(-m + \frac{1}{2}mv^2 \right) dt,$$

which, up to an irrelevant constant, gives the usual expression for a free particle Lagrangian as its kinetic energy.

In Chapter 2 we varied the path while keeping the metric fixed. Here, instead, we vary the metric:

$$\delta I_{\text{part}} = -\frac{m}{2} \int \left(g_{\kappa\lambda} \frac{dx^\kappa}{d\sigma} \frac{dx^\lambda}{d\sigma} \right)^{-1/2} \delta g_{\mu\nu} \frac{dx^\mu}{d\sigma} \frac{dx^\nu}{d\sigma} d\sigma = -\frac{m}{2} \int \delta g_{\mu\nu} \frac{dx^\mu}{ds} \frac{dx^\nu}{ds} ds, \tag{7.2}$$

where, as in Chapter 2, we choose $\sigma = s$ after the variation. This is not quite of the form (6.7), which involves an integral over spacetime, but we can make the expressions match by inserting a delta function, normalized so that

$$\int \delta^{(4)}(x-z) f(x) \sqrt{|g|}\, d^4x = f(z).$$

Then

$$\delta I_{\text{part}} = -\frac{m}{2} \iint \delta g_{\mu\nu} \frac{dz^\mu}{ds} \frac{dz^\nu}{ds} \delta^{(4)}(x - z(s)) \sqrt{|g|}\, ds\, d^4x, \tag{7.3}$$

where the particle trajectory is $x^\mu = z^\mu(s)$. We can now read off the stress-energy tensor from (6.7):

$$T^{\mu\nu} = m \int \frac{dz^\mu}{ds} \frac{dz^\nu}{ds} \delta^{(4)}(x - z(s))\, ds. \tag{7.4}$$

(There is an unobvious sign coming from the position of the indices of δg; see (4.28).)

In particular, in a flat spacetime with standard Cartesian coordinates, we can carrying out the remaining integral to obtain

$$T^\mu{}_0 = m \frac{dz^\mu}{ds} \delta^{(3)}(x^i - z^i(s)). \tag{7.5}$$

This may be recognized as the usual four-momentum density of a point particle, as anticipated in the preceding section.

7.3 Perfect fluids

For a collection of point particles, we can simply add stress-energy tensors of the form (7.4), each contributing along a particular world line. In the continuum limit, such a sum simplifies to

$$T^{\mu\nu} = \rho u^\mu u^\nu, \tag{7.6}$$

where the scalar ρ is a measure of energy density and $u^\mu(x)$ is the four-velocity dz^μ/ds of the fluid element at position x. As a next level of complexity, we can allow the particles to interact isotropically, that is, with no preferred direction. In the local rest frame of such a "perfect fluid," the components $T^0{}_i$ of stress-energy tensor must vanish; otherwise they would determine a preferred direction. For the same reason, the spatial components $T^i{}_j$ must have three equal eigenvalues; otherwise the eigenvectors would pick out preferred directions. Thus, in this frame,

$$T^\mu{}_\nu = \begin{pmatrix} \rho & 0 & 0 & 0 \\ 0 & -p & 0 & 0 \\ 0 & 0 & -p & 0 \\ 0 & 0 & 0 & -p \end{pmatrix}, \tag{7.7}$$

where physically, p is the fluid pressure. With our West Coast signature convention, this can be written in a frame-independent manner as

$$T^{\mu\nu} = (\rho + p)u^\mu u^\nu - pg^{\mu\nu}. \tag{7.8}$$

Normally one more equation is needed to specify the physics of a fluid. An equation of state gives a relationship between the pressure and the energy density, typically in the form $p = f(\rho)$. For many fluids, this equation of state is approximately linear,

$$p = w\rho, \tag{7.9}$$

where the "equation of state parameter" w is a number. For noninteracting particles—"dust" to cosmologists—$w = 0$. For radiation, momentum is equal to energy, but in a four-dimensional spacetime it is divided among three spatial directions, leading to an equation of state with $w = \frac{1}{3}$. A cosmological constant may be interpreted as a fluid with $w = -1$; with this choice, the stress-energy tensor is just $T^{\mu\nu} = \rho g^{\mu\nu}$, and the conservation equation forces ρ to be a constant. More exotic fluids can have other values of w, or equations of state that are more complicated than a linear relationship.

7.4 Other fields

For other fields, the stress-energy tensor can again be computed from the variational principle (6.7). To do so, of course, one needs to know the dependence of the action on the metric. This is usually determined by the "principle of minimal coupling": one starts with the special relativistic action, replaces the Minkowski metric with a general metric and partial derivatives with covariant derivatives, and inserts a factor of $\sqrt{|g|}$ in the integration measure. This procedure is not always unique, and there are cases in which a "nonminimal coupling" may be useful, but it is a good rule of thumb.

A scalar field φ with mass m and potential $V[\varphi]$, for instance, has an action

$$I_{\text{scalar}} = \int \left[\frac{1}{2}g^{\mu\nu}\nabla_\mu\varphi\nabla_\nu\varphi - \frac{1}{2}m^2\varphi^2 - V[\varphi]\right]\sqrt{|g|}\,d^4x. \tag{7.10}$$

(Here and elsewhere, signs depend on the metric signature convention; see Box 2.2.) Varying the metric, we can read off the stress-energy tensor

$$T^{\mu\nu}_{\text{scalar}} = \nabla^\mu\varphi\nabla^\nu\varphi - \frac{1}{2}g^{\mu\nu}(\nabla_\rho\varphi\nabla^\rho\varphi - m^2\varphi^2) + g^{\mu\nu}V. \tag{7.11}$$

In flat Minkowski space, this gives the expected energy density,

$$T^{00}_{\text{scalar}} = \frac{1}{2}\left(\dot\varphi^2 + |\nabla\varphi|^2 + m^2\varphi^2\right) + V. \tag{7.12}$$

An electromagnetic field is described by a potential A_μ and a field strength tensor

$$F_{\mu\nu} = \partial_\mu A_\nu - \partial_\nu A_\mu, \tag{7.13}$$

where in Minkowski space

$$F^{\mu\nu} = \begin{pmatrix} 0 & E_x & E_y & E_z \\ -E_x & 0 & B_z & -B_y \\ -E_y & -B_z & 0 & B_x \\ -E_z & B_y & -B_x & 0 \end{pmatrix}. \tag{7.14}$$

The electromagnetic action is

$$I_{\text{EM}} = -\frac{1}{4}\int F_{\mu\nu}F^{\mu\nu}\sqrt{|g|}\,d^4x = -\frac{1}{4}\int (F_{\mu\rho}F_{\nu\sigma}g^{\mu\nu}g^{\rho\sigma})\sqrt{|g|}\,d^4x, \tag{7.15}$$

from which we can read off

$$T^{\mu\nu}_{\text{EM}} = -F^{\mu\rho}F^\nu{}_\rho + \frac{1}{4}g^{\mu\nu}F_{\rho\sigma}F^{\rho\sigma}. \tag{7.16}$$

In flat Minkowski space, this gives

$$T^{00}_{\text{EM}} = \frac{1}{2}\left(|\mathbf{E}|^2 + |\mathbf{B}|^2\right), \quad T^{0i}_{\text{EM}} = (\mathbf{E}\times\mathbf{B})^i, \tag{7.17}$$

the usual electromagnetic energy density and Poynting vector.

7.5 Differential and integral conservation laws

Suppose that a vector J^μ in a flat spacetime satisfied a continuity equation

$$\partial_\mu J^\mu = 0, \tag{7.18}$$

as, for example, a current obtained from Noether's theorem (Box 6.3). Choosing a spatial region R with boundary ∂R, we can integrate at a fixed time to obtain

$$\frac{d}{dt}\int_R J^0\,d^3x = -\int_R \nabla\cdot\mathbf{J}\,d^3x = -\int_{\partial R}\mathbf{J}\cdot d\mathbf{S}. \tag{7.19}$$

This integral version is clearly a conservation law: it tells us that the change of a quantity in the region R is equal to the flux of a current through the boundary of R.

> **Box 7.1 Gravitational energy**
>
> There is a simple way to understand why the differential conservation law for the stress-energy tensor doesn't translate into an integral conservation law. As we have defined it, T^{ab} includes only the energy of matter, not gravitational energy. We know from Newtonian theory that this quantity cannot be conserved by itself; we have to add gravitational potential energy back in.
>
> The problem is that there is no local tensorial expression for gravitational energy density. At any spacetime point p, we can always choose a freely falling reference frame, in which the effects of gravity, and therefore any gravitational energy, disappear. But if a tensor is zero in one frame, it is zero in every frame. There has been a long search, going back to Einstein, for an "energy pseudotensor" for gravity, but it is not clear that any of the proposals really make sense. More recently, the most common approach has been to define quasilocal gravitational energy, energy in a finite region, in which no choice of frame can eliminate gravity.

In general relativity, a similar relation continues to hold if J^μ is a vector. By (5.28),

$$\nabla_\mu J^\mu = 0 \Rightarrow \partial_\mu(\sqrt{|g|} J^\mu) = 0, \tag{7.20}$$

and we can repeat the argument using $\sqrt{|g|} J^\mu$. For a higher rank tensor, though, this is no longer possible. In particular, the covariant derivatives in the conservation law (6.18) for the stress-energy tensor can't be written as ordinary partial derivatives—the connection terms get in the way—and we can't write the differential equation as an integral conservation law (see Box 7.1).

There is one interesting exception, though. Suppose the spacetime metric has a symmetry such as time translation invariance. As discussed in Section 4.6, this implies the existence of a Killing vector, a vector satisfying

$$\nabla_\mu \xi_\nu + \nabla_\nu \xi_\mu = 0. \tag{7.21}$$

Then

$$\nabla_\mu (T^{\mu\nu} \xi_\nu) = (\nabla_\mu T^{\mu\nu}) \xi_\nu + \frac{1}{2} T^{\mu\nu} (\nabla_\mu \xi_\nu + \nabla_\nu \xi_\mu) = 0. \tag{7.22}$$

The contraction $T^{\mu\nu} \xi_\nu$ is thus a conserved tangent vector, for which an integral conservation law of the form (7.19) exists. In particular, in a spacetime with a timelike Killing vector, we can choose coordinates such that $\xi^\mu = \delta^\mu_0$ (see Section A.7), and the four-momentum density $T^\mu{}_0$ obeys an integral conservation law.

7.6 Conservation and the geodesic equation

We finish this chapter with one of the most remarkable features of general relativity, the fact that the field equations determine the equations of motion of matter. Let us

> **Box 7.2 Field equations and equations of motion**
>
> Consider a set of massive particles, far enough apart that their gravitational interactions are weak. Enclose each in a small imaginary surface, which propagates in time to form a "world tube." Assume spacetime is empty outside the world tubes, so the vacuum Einstein field equations hold. Solve these equations, with boundary conditions that the metric at the edge of each world tube looks like a Schwarzschild metric.
>
> The result, as first found by Einstein, Infeld, and Hoffmann [63] in a slow motion approximation, is that the trajectories of the world tubes cannot be arbitrary. Rather, each particle must move along a geodesic in the background spacetime determined by the others. The results can be extended to include spin, higher multipole moments, and gravitational radiation reaction; the motion becomes more complicated, but it is still determined by the vacuum field equations.
>
> There is a nice way to check this result. Suppose we look for a solution of the Einstein field equations with two masses, and simply *demand* from the outset that the masses be stationary. Such a solution exists, but it has a singular "strut" along a line connecting the two masses. By considering parallel transport around this strut, we can compute its curvature and, through the Einstein field equations, its stress. The resulting stress has exactly the right value to physically hold the masses apart against their mutual attraction.

start with the simplest case, a fluid of noninteracting particles with the stress-energy tensor (7.6). We will need one simple identity:

$$u_\nu u^\nu = g_{\mu\nu} \frac{dx^\mu}{ds} \frac{dx^\nu}{ds} = 1 \Rightarrow u_\nu \nabla_\mu u^\nu = \frac{1}{2} \nabla_\mu (u_\nu u^\nu) = 0. \tag{7.23}$$

Now write out the conservation equation:

$$\nabla_\mu T^{\mu\nu} = 0 = \nabla_\mu (\rho u^\mu) u^\nu + \rho u^\mu \nabla_\mu u^\nu. \tag{7.24}$$

Multiplying by u_ν and using (7.23), we obtain

$$\nabla_\mu (\rho u^\mu) = 0. \tag{7.25}$$

This is a continuity equation of the type discussed in the preceding section; it implies conservation of mass for our fluid. Inserting this back into (7.24), we find

$$\rho u^\mu \nabla_\mu u^\nu = 0. \tag{7.26}$$

But this is just the geodesic equation! It tells us that wherever the fluid is present ($\rho \neq 0$), its particles follow geodesics. This was not put in by hand—the derivation of the stress-energy tensor in Section 7.2 assumed nothing about paths—but is rather a consequence of conservation, which is in turn implied by the Einstein field equations.

This result generalizes to other stress-energy tensors. For the perfect fluid (7.8), conservation gives the relativistic Euler equations, with the gradient of the pressure acting as a force. For charged particles, combining the particle stress-energy tensor (7.6) and the electromagnetic stress-energy tensor (7.16) gives the Lorentz force law. For the scalar stress-energy tensor (7.11),

$$\nabla_\mu T^{\mu\nu}_{\text{scalar}} = \left(\nabla_\mu \nabla^\mu \varphi + m^2 \varphi + \frac{dV}{d\varphi}\right) \nabla^\nu \varphi = 0. \tag{7.27}$$

The quantity in parentheses is the field equation for φ, so whenever $\nabla^\nu \varphi \neq 0$, conservation implies the field equation.

This phenomenon can be partially traced back to the derivation of the conservation law in Section 6.3. There, it was shown that diffeomorphism invariance implies stress-energy conservation, but only when the matter equations of motion are satisfied. Here we are basically reversing that argument.

But the result is deeper. As described in Box 7.2, the motion of matter can already be obtained from the *vacuum* field equations in the region outside the matter, without needing any details of the stress-energy tensor. This makes gravity very different from other interactions. Finding the motion of a charged particle in electrodynamics, for instance, takes two steps: first solving Maxwell's equations for the electromagnetic field, then using the Lorentz force law to find the acceleration in that field. These two steps must be independent, because the particle *won't* necessarily follow the path given by the Lorentz force law—there may be nonelectromagnetic forces acting as well, which would be invisible to Maxwell's equations.

What makes gravity special is its universality. Gravity sees *everything*; while other forces may be present, they will carry energy, and will be visible to the gravitational field. This universal coupling makes it possible to collapse the two steps into one, allowing the field equations to determine both the field and the motion.

Further reading

The discovery of the stress-energy tensor in the early 1900s was a complicated process; a piece of the history is chronicled in Section 9 of [58]. The definition of the stress-energy tensor given here is not unique; an alternative approach using Noether's theorem (Box 6.3) yields a tensor that can be slightly different. See [59] or, more briefly, Section 7.3 of [52] for discussions of their relationship. Chapter 4 of [10] and Chapter 6 of [60] offer deeper discussions of the stress-energy tensor of a fluid.

A discussion of gravitational energy pseudotensors can be found in Section 13.1 of [38]. There are several candidates for quasilocal gravitational energy. The idea became popular with the work of Brown and York [61]; for a review, see [62].

The demonstration that the Einstein field equations imply the equations of motion for sources originated in [63]. For a review of more recent developments, see [64–66]. The proof that two static masses must be joined by a "strut" with a stress that balances their gravitational attraction can be found in Chapter 8 of [67].

8
The weak field approximation

The Einstein field equations in four dimensions are ten coupled nonlinear partial differential equations. If the summations were written out in full, each equation would have hundreds of terms. Given this complexity, it is unlikely that we will find the general solution any time soon; we need other ways to extract physical implications. The next two chapters will look at a useful approximation, in which the metric is assumed to be only slightly different from that of flat Minkowski space.

8.1 The linearized field equations

Assume spacetime is "almost flat," in the sense of having a metric with components of the form
$$g_{\mu\nu} = \eta_{\mu\nu} + h_{\mu\nu} \tag{8.1}$$
in some suitable coordinate system, where η is the Minkowski metric (2.26) and h and its derivatives are assumed to be small. As we can see from the Schwarzschild metric of Chapter 3, this is a not an unreasonable approximation—in the Solar System, even just outside the Sun, h is at most about 4×10^{-6}. The inverse metric is
$$g^{\mu\nu} = \eta^{\mu\nu} - h^{\mu\nu} + \mathcal{O}(h^2), \tag{8.2}$$
while the determinant is
$$|g| = 1 + \eta^{\mu\nu} h_{\mu\nu} + \mathcal{O}(h^2). \tag{8.3}$$
In this chapter and the next, we will raise and lower indices with η rather than g, as is conventional in the weak field approximation.

To first order in h, the Levi-Civita connection (5.23) now becomes
$$\Gamma^{\rho}_{\mu\nu} = \frac{1}{2}\eta^{\rho\sigma}\left(\partial_\mu h_{\nu\sigma} + \partial_\nu h_{\mu\sigma} - \partial_\sigma h_{\mu\nu}\right), \tag{8.4}$$
and the Ricci tensor (6.10) is
$$R_{\mu\nu} = \frac{1}{2}\left(\partial_\mu\partial^\sigma h_{\nu\sigma} + \partial_\nu\partial^\sigma h_{\mu\sigma} - \Box h_{\mu\nu} - \partial_\mu\partial_\nu h\right), \tag{8.5}$$
where
$$h = \eta^{\rho\sigma} h_{\rho\sigma} \quad \text{and} \quad \Box = \eta^{\rho\sigma}\partial_\rho\partial_\sigma. \tag{8.6}$$
To write down the Einstein tensor, one more abbreviation is useful. Define

$$\bar{h}_{\mu\nu} = h_{\mu\nu} - \frac{1}{2}\eta_{\mu\nu}h. \tag{8.7}$$

This quantity is sometimes called the "trace reversal" of $h_{\mu\nu}$, since

$$\bar{h} = \eta^{\sigma\tau}\bar{h}_{\sigma\tau} = -h \;\Rightarrow\; h_{\mu\nu} = \bar{h}_{\mu\nu} - \frac{1}{2}\eta_{\mu\nu}\bar{h}. \tag{8.8}$$

A simple calculation then shows that to first order in h, the Einstein tensor is

$$G_{\mu\nu} = \frac{1}{2}\left(-\Box\bar{h}_{\mu\nu} + \partial_\mu\partial^\sigma\bar{h}_{\nu\sigma} + \partial_\nu\partial^\sigma\bar{h}_{\mu\sigma} - \eta_{\mu\nu}\partial^\rho\partial^\sigma\bar{h}_{\rho\sigma}\right). \tag{8.9}$$

We can further simplify this expression through a coordinate choice. Our coordinate freedom is now somewhat limited: the splitting (8.1) of the metric into a "background" and a "perturbation" is coordinate-dependent, and we have to restrict ourselves to coordinates that preserve that splitting. But a transformation $x^\mu \to x^\mu + \xi^\mu$ is still permissible as long as ξ^μ is of order h. Under such a transformation, we know from Section 4.6 that

$$\delta g_{\mu\nu} = \nabla_\mu \xi_\nu + \nabla_\nu \xi_\mu = \partial_\mu \xi_\nu + \partial_\nu \xi_\mu + \mathcal{O}(h^2) \;\Rightarrow\; h_{\mu\nu} \to h_{\mu\nu} + \partial_\mu \xi_\nu + \partial_\nu \xi_\mu, \tag{8.10}$$

and hence, by an easy calculation,

$$\partial^\sigma \bar{h}_{\mu\sigma} \to \partial^\sigma \bar{h}_{\mu\sigma} + \Box \xi_\mu. \tag{8.11}$$

If $\partial^\sigma \bar{h}_{\mu\sigma}$ is nonzero in our original coordinate system, we can now choose ξ to satisfy $\Box \xi_\mu = -\partial^\sigma \bar{h}_{\mu\sigma}$. Then in our new coordinates,

$$\partial^\sigma \bar{h}_{\mu\sigma} = 0 \tag{8.12}$$

and thus

$$G_{\mu\nu} = -\frac{1}{2}\Box\bar{h}_{\mu\nu} + \mathcal{O}(h^2). \tag{8.13}$$

The coordinate choice (8.12) is called de Donder gauge, harmonic gauge, Fock gauge, Lorenz gauge (after Ludvig Lorenz, not the Hendrik Lorentz of special relativity), or de Donder coordinates, harmonic coordinates, Fock coordinates, or various combinations of these names. It is the first-order approximation of a more general coordinate condition,

$$\frac{1}{\sqrt{|g|}}\partial_\mu\left(\sqrt{|g|}g^{\mu\rho}\right) = g^{\mu\nu}\Gamma^\rho_{\mu\nu} = 0, \tag{8.14}$$

which was used by Einstein in a preliminary version of general relativity. In these coordinates, the Einstein field equations with $\Lambda = 0$ reduce to an ordinary wave equation,

$$\Box \bar{h}_{\mu\nu} = -2\kappa^2 T_{\mu\nu} + \mathcal{O}(h^2). \tag{8.15}$$

8.2 The Newtonian limit

Let us next make a further approximation, an assumption that all objects are moving at speeds much slower than light. For the stress-energy tensor (7.6), this means

$$T_{00} \approx \rho \quad \text{with all other components vanishing.} \tag{8.16}$$

The same is true for the perfect fluid stress-energy tensor (7.8), since the pressure p is really p/c^2 in units $c \neq 1$. The d'Alembertian $\Box$ with factors of c restored is

$$\Box = \frac{1}{c^2}\frac{\partial^2}{\partial t^2} - \nabla^2,$$

so we can neglect the time derivatives. The weak field equation (8.15) then reduces to

$$\nabla^2 \bar{h}_{00} = 2\kappa^2 \rho. \tag{8.17}$$

The other components of $\bar{h}_{\mu\nu}$ obey a Laplace equation with no source, so with any reasonable fall-off conditions at infinity, they must be zero.

Comparing to the Poisson equation for the Newtonian gravitational potential Φ,

$$\nabla^2 \Phi = 4\pi G \rho,$$

we see that

$$\bar{h}_{00} = \alpha \Phi \quad \text{with} \quad \alpha = \frac{\kappa^2}{2\pi G}. \tag{8.18}$$

We can determine the original, non-trace-reversed perturbation $h_{\mu\nu}$ from (8.8):

$$h_{00} = \frac{\alpha}{2}\Phi, \quad h_{0i} = 0, \quad h_{ij} = \frac{\alpha}{2}\Phi \delta_{ij}, \tag{8.19}$$

and thus

$$ds^2 = \left(1 + \frac{\alpha}{2}\Phi\right) dt^2 - \left(1 - \frac{\alpha}{2}\Phi\right)(dx^2 + dy^2 + dz^2). \tag{8.20}$$

This is exactly the line element whose geodesics we investigated in Section 2.5. There, we saw that it gave the correct Newtonian equations of motion provided

$$\frac{\alpha}{2} = 2 \Rightarrow \kappa^2 = 8\pi G, \tag{8.21}$$

in agreement with the result (6.17) coming from the geodesic deviation equation.

Observationally, this result might be called the "zeroth test" of general relativity. The classical tests derived in Chapter 3 were crucial in establishing the theory, but without this Newtonian limit, they would have been irrelevant: there's no point in predicting corrections to Newtonian gravity if you can't reproduce the Newtonian predictions to start with.

8.3 Gravitomagnetism

For the Newtonian approximation, we neglected all terms of order v/c or higher. Let us now sketch out what happens at the next order.

First, we will need to keep the terms $T^{0i} = \rho u^i$ in the stress-energy tensor. This means that in addition to $\bar{h}_{00}$ we now have a nontrivial $\bar{h}_{0i} = h_{0i}$, with

$$\nabla^2 h_0{}^i = 16\pi G \rho u^i. \tag{8.22}$$

The solution will depend on the mass and velocity distributions, but we can gain a quick intuition by comparing with the electromagnetic potential (again to order v/c):

Gravity	Electromagnetism	
$\nabla^2 \Phi = 4\pi G \rho$	$\nabla^2 \phi_{\text{EM}} = -4\pi \rho_{\text{charge}}$	
$\nabla^2 h_0{}^i = 16\pi G \rho u^i$	$\nabla^2 A^i_{\text{EM}} = 4\pi \rho_{\text{charge}} u^i$	(8.23)

where ρ_{charge} is the charge density. The sign difference in the potential comes from the fact that like charges repel while like masses attract, and the extra factor of four in the vector equation reflects the different tensorial structures. Otherwise, though, the equations are exact analogs; if we can find the electromagnetic potential from a current density, we can find the metric perturbation from the corresponding mass current density.

By itself, this would seem to merely be a calculational shortcut. In fact, it's more. To first order in v/c, the geodesic equation is

$$\frac{d^2 x^i}{dt^2} + \Gamma^i_{00} + 2\Gamma^i_{0j}\frac{dx^j}{dt} = 0 \;\Rightarrow\; \delta_{ij}\frac{d^2 x^j}{dt^2} = -\partial_i \Phi + (\partial_i h_{0j} - \partial_j h_{0i})\frac{dx^j}{dt}. \tag{8.24}$$

This has exactly the same structure as the Lorentz force law in electromagnetism. Just as a charge current creates a magnetic field that exerts a $\mathbf{v} \times \mathbf{B}$ force, a mass current creates a "gravitomagnetic" field that exerts what looks like a $\mathbf{v} \times \mathbf{B}_{\text{grav}}$ force. This phenomenon is also called the Lense-Thirring effect, after its discoverers [68], or frame-dragging, since a spinning mass will "drag" objects in the direction of its rotation.

Gravity is much weaker than electromagnetism, and gravitomagnetism is not easy to detect. It has now been observed, though, both in its effect on satellite orbits—in particular, through measurements of the LAGEOS satellites—and in the precession of the extremely stable and accurate gyroscopes of the Gravity Probe B mission.

8.4 Higher orders

We ended the expansion in Section 8.1 at first order in h. With more work, we can continue on to higher orders, to obtain a series expansion for solutions of the field equations. Such an expansion is called the post-Newtonian expansion if velocities are also assumed to be small, or the post-Minkowskian expansion otherwise.

At the next order, the connection has new terms with the structure $h\partial h$, and the Einstein tensor picks up terms of the form $(\partial h)^2$. Moved to the right-hand side

Box 8.1 Gravity gravitates

At second order in the weak field expansion, the field equations have the form

$$\Box \bar{h}_{\mu\nu} = -2\kappa^2 (T_{\mu\nu} + t_{\mu\nu}), \quad (8.25)$$

where $\kappa^2 t \sim (\partial h)^2$ is a quadratic function of h and its derivatives. Although $t_{\mu\nu}$ is not a tensor (see Box 7.1), it can be loosely interpreted as gravitational energy, acting as a source for the gravitational field.

This phenomenon can be tested through its effect on the Moon's orbit. The inertial masses of the Earth and the Moon each have a small component E_{grav}/c^2 coming from gravitational binding energy. Because of differences in composition, though, this energy contributes a different proportion of the total mass of the two bodies. If this binding energy did not itself gravitate, the equivalence principle would be violated, and the Earth and Moon would orbit the Sun with slightly different accelerations, a phenomenon called the Nordtvedt effect [69]. Thanks to Lunar laser ranging, the position of the Moon is known to millimeter precision, and a nonzero Nordtvedt effect has been ruled out to a few parts in 10^4.

---

of the field equations (8.15), these terms form an effective stress-energy tensor for the gravitational field. This result—the fact that gravitational energy gravitates—has been tested to surprisingly good accuracy, as described in Box 8.1. At higher orders, the expansion incorporates such added effects as gravitational radiation reaction and spin-orbit and spin-spin couplings.

We can also expand around a curved background metric, $g_{ab} = \tilde{g}_{ab} + h_{ab}$. Define

$$h = \tilde{g}^{ab} h_{ab}, \quad \bar{h}_{ab} = h_{ab} - \frac{1}{2} \tilde{g}_{ab} h, \quad (8.26)$$

where we now raise and lower indices with $\tilde{g}$. Choose the coordinate condition

$$\tilde{\nabla}_a \bar{h}^{ab} = 0, \quad (8.27)$$

where $\tilde{\nabla}$ is the covariant derivative compatible with the background metric $\tilde{g}$. After a bit of calculation, the first order field equations take the form

$$\widetilde{\Box} \bar{h}_{ab} + \tilde{R}_a{}^c \bar{h}_{bc} + \tilde{R}_b{}^c \bar{h}_{ac} + 2\tilde{R}_{acdb} \bar{h}^{cd} = -2\kappa^2 T_{ab}. \quad (8.28)$$

The differential operator on the left-hand side is known as the Lichnerowicz operator. This form is useful for studying, for example, small perturbations of a black hole metric such as the "quasinormal modes" that are generated as the system settles into equilibrium, as well as wave propagation in a cosmological background.

8.5 Bootstrapping the field equations

There is nothing to stop us from carrying out the weak field expansion to arbitrarily high order. But we can also reverse the process, and obtain a very different derivation of the Einstein field equations.

Let us briefly forget curved spacetime, and try to construct a theory of gravity in the flat spacetime of special relativity. To obtain an inverse square law interaction, we need a massless field. For this field to have the stress-energy tensor as its source, it must be represented by a symmetric tensor $\bar{h}_{\mu\nu}$, and should, at least to first order, obey a wave equation of the form (8.15). To ensure exact masslessness, we must also impose a symmetry (or "gauge invariance") of the form (8.10). In the language of quantum field theory, such an object is a massless spin two field.

It's easy to write down an action $I^{(1)}$ that gives the field equations (8.15). But this is not quite enough: for consistency, the right-hand side of the field equations should include the stress-energy tensor of the field $\bar{h}_{\mu\nu}$ itself. We can compute this stress-energy tensor from $I^{(1)}$, as we would for any other field, and include it as a source (see Box 8.1). But to obtain *these* field equations from an action, we need an additional term $I^{(2)} \sim h(\partial h)^2$. This new term, in turn, modifies the stress-energy tensor of $\bar{h}_{\mu\nu}$, requiring a further $\mathcal{O}(h^3)$ correction to the right-hand side of the field equations. This correction then requires a new term $I^{(3)} \sim h^2(\partial h)^2$ in the action.

It seems that we need an infinite iteration. With a clever choice of variables, though, it is possible to find a series that terminates after finitely many terms. The resulting action turns out to be exactly the standard Einstein-Hilbert action, written in terms of a metric $g_{\mu\nu} = \eta_{\mu\nu} + h_{\mu\nu}$. As we saw in Chapter 7, this means that the sources of the gravitational field must move along geodesics of this metric, appropriately modified by any nongravitational forces that might be present. The flat metric $\eta_{\mu\nu}$ we started with has completely vanished as an independent entity. So even if we start with flat spacetime, self-consistency and the universality of the gravitational coupling force us into a setting that is indistinguishable from a curved spacetime.

Further reading

The weak field approximation already appeared in a 1916 paper by Einstein [70], in which he used it to predict the existence of gravitational waves. The de Donder gauge used here is not the only convenient choice; see Chapter 7 of Carroll's *Spacetime and Geometry* [17] for other useful coordinate choices.

A thorough discussion of gravitomagnetism can be found in Ciufolini and Wheeler's *Gravitation and Inertia* [71]. Observational results for both gravitomagnetism and the Nordtvedt effect are described in [4].

Second order terms in the weak field expansion are derived in Chapter 35 of [11]. Higher order terms are a work in progress; see [64–66] for some results. The weak field expansion in a curved background described here—the "short wave approximation"—was first studied by Isaacson [72]. Details can be found in Chapter 35 of [11].

The derivation of the Einstein-Hilbert action from flat spacetime has a long history, culminating in a paper by Deser [73]. An interesting description from a quantum field theoretical perspective can be found in the *Feynman Lectures on Gravitation* [74].

9
Gravitational waves

As we saw in the preceding chapter, small fluctuations in the metric are governed by a wave equation. In this chapter, we will explore the solutions to this equation, gravitational waves, as well as some of the physics involved in their detection. Gravitational wave astronomy is a new and rapidly growing field; the aim here is not to offer a comprehensive description, but to provide tools to allow further study.

We can gain some initial intuition for gravitational waves with a quasi-Newtonian argument. As we know from electromagnetism, a multipole expansion can be a useful tool for describing wave behavior. As a source of gravitational waves, consider a clump of particles with masses m_i and positions $\mathbf{x}_i$, and with a center of mass position $\bar{\mathbf{x}}$. In a multipole expansion,

- The gravitational monopole moment is $M = \sum_i m_i$,
  which is constant by mass conservation.
- The gravitational dipole moment is $\mathbf{d} = \sum_i m_i (\mathbf{x}_i - \bar{\mathbf{x}})$,
  which is constant by momentum conservation.
- The gravitational "magnetic dipole moment" is $\mathbf{L} = \sum_i m_i \mathbf{v}_i \times \mathbf{x}_i$,
  which is constant by angular momentum conservation.

We might thus expect the first appearance of gravitational radiation to be quadrupolar. As we shall see shortly, this is correct.

9.1 Solving the weak field equations

In a curved background, the weak field equations (8.28) involve a curvature term, which can be rather complicated. In a flat background, though, eqn (8.15) is exactly the same as Maxwell's equations for the electromagnetic potential in Lorenz gauge, and the solution can be simply copied from electrodynamics. In direct analogy to the Liénard-Wiechert solution, we have

$$\bar{h}_{\mu\nu}(t,\mathbf{x}) = 4G \int \frac{T_{\mu\nu}(t - |\mathbf{x}-\mathbf{y}|, \mathbf{y})}{|\mathbf{x}-\mathbf{y}|} d^3\mathbf{y}. \tag{9.1}$$

As in electrodynamics, the source—here, the stress-energy tensor—appears in this expression at retarded time $t - |\mathbf{x}-\mathbf{y}|$. This is a way of saying that waves travel at the speed of light: the perturbation $\bar{h}_{\mu\nu}$ at time t at a distance L from the source is determined by the source at the earlier time $t - L/c$.

We will focus on the purely spatial components $\bar{h}_{ij}$ ($i,j = 1,2,3$) of the perturbation, since the de Donder coordinate condition (8.12) then determines the remaining

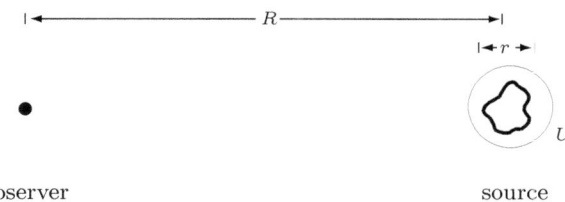

Fig. 9.1 Gravitational waves from a distant source

components. Suppose the source is confined to a small region of size r, which we observe at a much larger distance $R \gg r$, as shown in Fig. 9.1. We can then approximate the denominator in (9.1) as $|\mathbf{x} - \mathbf{y}| \approx R$, so

$$\bar{h}_{ij}(t, \mathbf{x}) \approx \frac{4G}{R} \int_U T_{ij}(t - R, \mathbf{y})\, d^3y, \qquad (9.2)$$

where the integral is over any region U that contains the source. We can now use the conservation law, which at this order tells us that

$$\partial_\mu T^{\mu\nu} = 0 \Rightarrow \partial_0 T^{00} = -\partial_k T^{k0}, \quad \partial_0 T^{0i} = -\partial_k T^{ki}. \qquad (9.3)$$

Define

$$\mathcal{Q}^{ij} = x^i x^j T^{00}. \qquad (9.4)$$

Taking one time derivative,

$$\partial_0 \mathcal{Q}^{ij} = x^i x^j \partial_0 T^{00} = -x^i x^j \partial_k T^{k0} = -\partial_k \left(x^i x^j T^{k0} \right) + x^j T^{i0} + x^i T^{j0}, \qquad (9.5)$$

since $\partial_k x^i = \delta_k^i$. A second time derivative then gives

$$\partial_0^2 \mathcal{Q}^{ij} = -\partial_k \left(x^i x^j \partial_0 T^{k0} \right) - x^j \partial_k T^{ik} - x^i \partial_k T^{jk}$$
$$= -\partial_k \left(x^i x^j \partial_0 T^{k0} + x^j T^{ik} + x^i T^{jk} \right) + 2T^{ij}. \qquad (9.6)$$

If we integrate the right-hand side of (9.6) over the region U, the first term becomes a surface integral, which vanishes because $T^{\mu\nu} = 0$ at the boundary. So substituting into (9.2), we finally obtain

$$\bar{h}_{ij}(t, \mathbf{x}) \approx \frac{2G}{R} \int_U \partial_0^2 \mathcal{Q}_{ij}(t - R, \mathbf{y})\, d^3y = \frac{2G}{R} \frac{d^2}{dt^2} \int_U y_i y_j T^{00}(t - R, \mathbf{y})\, d^3y. \qquad (9.7)$$

The integral on the right-hand side of this equation is the quadrupole moment of the source. As anticipated, the leading order gravitational radiation is quadrupolar.

We can now obtain a rough estimate of radiated power. Consider a gravitationally bound system of a few bodies with masses of order m, moving at speeds much less

> **Box 9.1 Conventions: part VI**
>
> The integral
> $$I_{ij} = \int_U y_i y_j T^{00}(t, \mathbf{y}) d^3\mathbf{y}$$
> is the second moment of the mass distribution. It is also known as the quadrupole moment. But the term "quadrupole moment" sometimes refers instead to the traceless part of I_{ij}, or three times the traceless part, or $3/2$ times the traceless part. Again, one must carefully check conventions.

than c. From Newtonian gravity, typical velocities and accelerations are $v^2 \sim Gm/r$, $a \sim Gm/r^2$, so from (9.7),

$$h \sim \frac{Gm}{R} v^2, \quad \dot{h} \sim \frac{Gm}{R} va \sim \frac{1}{R}\left(\frac{Gm}{r}\right)^{5/2}. \tag{9.8}$$

As described in Box 8.1, the effective gravitational stress-energy tensor is

$$t \sim \frac{1}{G}\dot{h}^2 \sim \frac{1}{R^2}\frac{G^4 m^5}{r^5}.$$

The total power is obtained by integrating over a sphere of radius R,

$$P \sim \frac{G^4 m^5}{r^5 c^5}, \tag{9.9}$$

where the factors of c have been restored by dimensional analysis. This agrees with the exact calculation up to factors of order one.

For typical stellar systems, this power is tiny. The Solar System, for example, radiates about 5 kW in gravitational waves. A binary neutron star system, on the other hand, can emit 10^{25} W, and a merging stellar-mass black hole binary can have a peak emission of more than 10^{49} W. The first observational confirmation of gravitational radiation came from the "radiation reaction" of the Hulse-Taylor binary pulsar, the orbital decay of the system caused by this energy loss.

9.2 Propagating waves

Equation (9.7) describes the production of gravitational waves. Let us now turn to their subsequent propagation. This may seem trivial—the waves obey an elementary wave equation $\Box \bar{h}_{\mu\nu} = 0$—but an extra coordinate symmetry makes the description even simpler.

Recall that in Chapter 8, we chose coordinates to impose the supplementary de Donder condition $\partial^\sigma \bar{h}_{\mu\sigma} = 0$. This does not quite use up our coordinate freedom,

Box 9.2 The speed of gravity

It is clear from the wave equation (8.15) that perturbations of a flat metric move at the speed of light. Though it's less obvious, (8.28) implies the same for any background. Beyond perturbation theory, though, the issue becomes tricky: the "speed of light" is determined by null geodesics, which change when the gravitational field changes. But by rephrasing the question as an initial value problem, it can be shown that arbitrary changes in initial data in a small region propagate outward at light speed. This prediction has now been confirmed to very high accuracy by the simultaneous observation of electromagnetic and gravitational waves from merging neutron stars.

But this result also presents a puzzle. Picture two identical stars orbiting each other. If Newtonian gravity propagated at a finite speed, the force on each star would point toward the retarded position of the other, not quite toward the center of mass, and the orbit would become unstable. As early as 1805, Laplace noted that observations of the Moon's orbit implied a speed of (Newtonian) gravity of at least $7 \times 10^6\, c$.

In general relativity, though, the gravitational field of a moving object depends not only on its position, but also on its velocity and acceleration. One can show that the resulting velocity-dependent interactions cancel out most of the effect of time delay. The effective force points not toward the retarded position, but toward the retarded position "extrapolated forward" to very nearly the instantaneous position, in a way that is consistent with both causality and stability.

---

though—de Donder gauge will be preserved under further transformations (8.10) as long as
$$\Box \xi^\mu = 0.$$
This is a rather strong restriction, but if the perturbation already satisfies a sourceless wave equation, it's not so bad: if we transform coordinates at an initial time, the transformation will propagate along with the fields. In particular, it may be shown that for a wave propagating in empty space, we can consistently choose "radiation gauge,"
$$h_{\mu 0} = 0, \quad h = 0. \tag{9.10}$$
A plane wave in radiation gauge will take the form
$$h_{ij} = \varepsilon_{ij} \sin(k_\rho x^\rho) \quad \text{with} \quad k_\mu k^\mu = 0, \quad k^i \varepsilon_{ij} = 0, \quad \varepsilon^i{}_i = 0. \tag{9.11}$$
Two polarizations remain: for propagation in the z direction, for instance,
$$\varepsilon_{ij} = \begin{pmatrix} \varepsilon_+ & \varepsilon_\times & 0 \\ \varepsilon_\times & -\varepsilon_+ & 0 \\ 0 & 0 & 0 \end{pmatrix}. \tag{9.12}$$

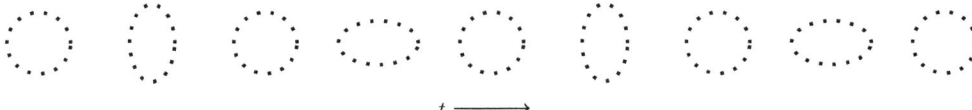

Fig. 9.2 A ring of test particles stretched and squeezed as a gravitational wave passes

The + and × polarizations differ by a 45° rotation; in quantum field theory, this means the excitations are spin two. Other models of gravity can allow additional polarizations, so polarization measurements offer an important test of general relativity.

9.3 Detecting gravitational waves

Imagine a weak gravitational plane wave with a + polarization moving along the z axis. (The × polarization will be identical except for a 45° rotation.) From the preceding section, the line element will be, to first order,

$$ds^2 = dt^2 - dz^2 - (1 - \varepsilon_+ \sin \omega(t-z))\, dx^2 - (1 + \varepsilon_+ \sin \omega(t-z))\, dy^2. \quad (9.13)$$

Choose these coordinates to define a "rest frame," keeping in mind that statements about "rest" are coordinate-dependent. Now consider a test particle—an object small enough that its own gravitational field can be ignored—at rest at some initial moment $t = 0$. The geodesic equations give

$$\left[\frac{d^2 x^i}{ds^2} + \Gamma^i_{\mu\nu} \frac{dx^\mu}{ds}\frac{dx^\nu}{ds}\right]_{t=0} = \frac{d^2 x^i}{ds^2}\bigg|_{t=0} = 0, \quad (9.14)$$

since $\Gamma^i_{00} = 0$ for the metric (9.13). Hence the object will remain at rest in this frame.

But the *proper* distance between two test particles at a fixed time, the physically measured distance, is not the coordinate distance. Rather, it is $\int d\ell$, where

$$d\ell^2 = dz^2 + (1 - \varepsilon_+ \sin \omega(t-z))\, dx^2 + (1 + \varepsilon_+ \sin \omega(t-z))\, dy^2. \quad (9.15)$$

In particular, if two test particles lie on a plane of constant $z = z_0$ with coordinate distances $\Delta x = \ell_0 \cos\theta$ and $\Delta y = \ell_0 \sin\theta$, their physical distance will be

$$\ell = \ell_0 \left[1 - \frac{1}{2}\varepsilon_+ \cos 2\theta \sin \omega(t-z_0)\right] + \mathcal{O}(\epsilon_+^2). \quad (9.16)$$

A ring of test particles will be distorted in the pattern shown in Fig. 9.2, as perpendicular directions alternately stretch and shrink. The light travel time between two particles is just $dt = d\ell$, so it, too, will oscillate.

We can confirm this picture in a more clearly coordinate-independent way by looking at the weak field geodesic deviation equation. To first order in h, (6.3) now becomes

$$\frac{d^2 X^i}{dt^2} = \frac{1}{2}\left(\partial_0^2 h^i{}_k\right) X^k. \quad (9.17)$$

For small fluctuations around an initial separation X_0^i, this implies

> **Box 9.3 Strong fields**
>
> For systems like merging black holes or neutron stars, the weak field approximation breaks down near the merger, where deviations from flat spacetime become large. In some cases, other approximations may be available—for example, for an "extreme mass ratio" inspiral in which the ratio of the two masses provides a new small parameter. In general, though, one needs to solve the Einstein field equations on a computer to obtain reliable predictions. Numerical relativity, especially in the context of gravitational waves, is a major field of research, but, unfortunately, one that does not fit well into an introductory textbook. See Section 13.1 for a bit more discussion and further references.

$$X^i = \left(\delta^i{}_k + \frac{1}{2}h^i{}_k\right) X_0^k, \tag{9.18}$$

which agrees with (9.16).

Physically, h represents a strain, a relative change in length, and a gravitational wave produces an oscillating strain. This pattern of systematic distortion of physical distance is the basis for the detection of gravitational waves. The first attempts, by Joseph Weber, looked for strain induced in a large aluminum cylinder. Subsequent resonant bar detectors have become more sensitive, but at this writing they have not yet detected any signals.

More sensitive modern detectors use laser interferometry. The distortions shown in Fig. 9.2 will lengthen one arm of a Michelson interferometer while shortening the other, causing a shift in interference fringes. The first direct detection of a gravitational wave was made by the LIGO observatory in 2015. Simultaneous observations in two widely separated interferometers showed a wave pattern that very closely matched the predictions of general relativity for the inspiral, merger, and subsequent quasinormal mode "ringdown" of two black holes.

9.4 Sources and signals

The detection of gravitational waves, either with resonant bars or with laser interferometers, is a direct detection of the strain h. From (9.8), restoring factors of c,

$$h \sim \left(\frac{r}{R}\right)\left(\frac{G}{c^4}\right)\left(\frac{mv^2}{r}\right). \tag{9.19}$$

The third term is essentially the centripetal force required to hold the radiating system together. The second term is the inverse of the "Planck force," $c^4/G \approx 1.2 \times 10^{44}$ N.

Laboratory generation of detectable signals is thus clearly out of reach. Next generation detectors may be sensitive to strains as small as $h \sim 10^{-25}$, but to achieve even that value, a spinning rod with a $1\,\text{m}^2$ cross-section would need a tensile strength of 10^{10} Gpa, more than nine orders of magnitude larger than the tensile strength of carbon fiber.

78   *Gravitational waves*

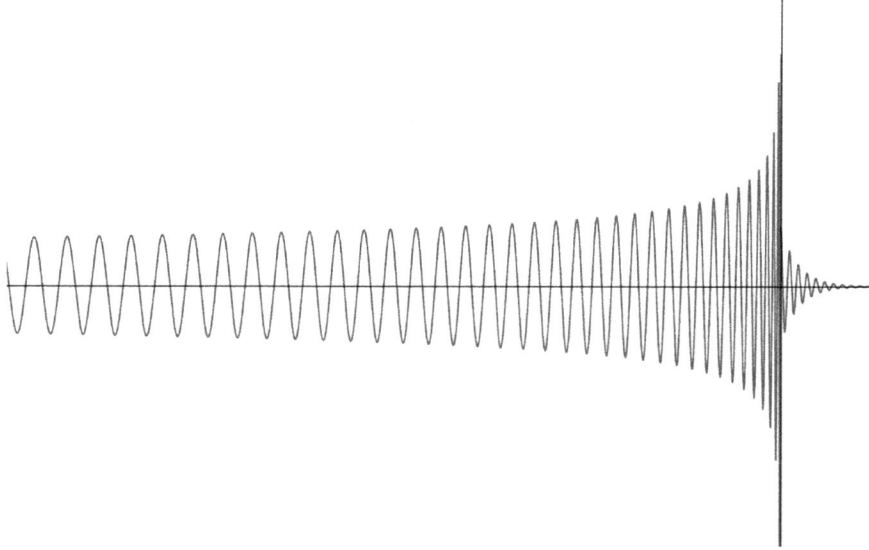

Fig. 9.3 A sketch of the characteristic "chirp" wave form of a binary system

Astrophysical sources, on the other hand, can generate much higher strains. These sources are much farther away, of course, but fortunately we directly measure h, which falls off as $1/R$, rather than energy, which falls off as $1/R^2$. Some plausible sources include:

- **Compact binaries:**

    A pair of orbiting massive objects—in particular, black holes or neutron stars—emits gravitational radiation. As the system radiates away energy, the bodies spiral inward, decreasing the distance r and increasing the velocity v. In the final stages of inspiral and merger, gravitational radiation can become strong enough to be observed at large distances. The first detected black hole binary, for instance, was at a distance of about 440 Mpc. The radiation has a characteristic "chirp" pattern, as illustrated in Fig. 9.3: as the system spirals inward, both the amplitude and the frequency increase, with a rapid rise just before merger. If the end state is a black hole, there is also a characteristic "ringdown" of quasinormal modes as it settles into its final stationary state.

- **Rotating neutron stars:**

    If a rotating neutron star is exactly axially symmetric, its quadrupole moment is constant, and it will emit no gravitational radiation. A deformation of the surface, though, can lead to a large and rapidly changing quadrupole moment—picture a "lump" on a spinning ball—and could be a substantial source of radiation. The observation of gravitational radiation from a known neutron star would serve as a measurement of its shape; nonobservation already places limits on asphericity.

- **Supernovae:**

    A supernova generates an enormous amount of energy. As with a neutron star, though, a perfectly spherical collapse and explosion will generate no gravitational waves. We don't yet understand the internal geometry of the supernova process well enough to reliably estimate the production of gravitational radiation.

- **Primordial gravitational radiation:**

    Just as the early Universe generated primordial electromagnetic radiation, the cosmic microwave background (see Section 11.3), it should have also generated a primordial gravitational wave background. A stochastic background of primordial gravitational waves can potentially be detected through its effect on the polarization of the microwave background. The amount of this radiation is highly sensitive to the details of cosmology of the very early Universe, though, and we do not yet have reliable predictions.

- **Unknown sources:**

    In the past, every time we have discovered a new way to observe the Universe, from the telescope to X-ray astronomy, we have found new, unexpected sources. It seems plausible that the emerging field of gravitational wave astronomy will do the same.

Further reading

The quadrupole formula was already present in Einstein's 1916 paper on weak fields [70]. For a proof of the existence of the radiation gauge (9.10), see for example Section 4.4 of Wald's textbook, *General Relativity* [35]. Maggiore's *Gravitational Waves* gives a broad introduction to gravitational radiation [66].

The waveform of Fig. 9.3 is an approximation based on [75]. References for strong field computations are given at the end of Chapter 13. The demonstration that arbitrarily strong perturbations of the gravitational field propagate at light speed may be found in [76]. For a discussion of the resulting puzzle described in Box 9.2, see [77].

Weber's first attempts to detect gravitational waves are discussed in [78]. The first indirect detection of gravitational radiation, from the energy loss and orbital decay of a binary pulsar, is described in [79, 80]. The first direct observation, by the LIGO interferometers, was reported in [81]. Both discoveries led to Nobel Prizes, the former for Hulse and Taylor, the latter for Weiss, Barish, and Thorne.

Gravitational wave physics is a rapidly developing field, and "current" references here may be out of date within a few years. Two good sources for reasonably up-to-date information are the journal *Living Reviews in Relativity*, http://www.springer.com/us/livingreviews, which has a section of reviews of gravitational wave physics, and the LIGO Scientific Collaboration, https://www.ligo.org.

10
Black holes

In Chapter 3, we used the Schwarzschild solution to study the spacetime geometry of the Solar System. It's now time to derive that metric, and to look more carefully at its properties. In particular, we have not yet explored the puzzling region near $r = 2m$, where the standard Schwarzschild metric seems to blow up. This region was irrelevant for Chapter 3: for the Solar System, it would be deep inside the Sun, where the vacuum field equations no longer hold. But if, instead, we assume "vacuum all the way down," we will see that the Schwarzschild metric can be reinterpreted as the spacetime of a black hole.

10.1 Static spacetimes

The two preceding chapters took advantage of the weak field approximation to simplify the Einstein field equations to a system we could actually solve. There is another way to gain control over the equations: we can search for spacetimes with exact symmetries. If we impose enough symmetry, the number of independent variables and equations can be reduced to the point that we can find exact solutions.

We know from Section 4.6 that a symmetry of the metric, an isometry, reveals itself through the presence of a Killing vector. Properly speaking, we should impose symmetries by requiring the existence of Killing vectors with certain properties. For now, we will take a shortcut, and describe symmetries in a coordinate-dependent manner; a more purely geometric approach is given in Section A.7.

The Schwarzschild metric (3.1) is time-independent, or "stationary." Geometrically, this means there is a timelike Killing vector ξ, that is, a timelike direction along which the metric is constant. As one might expect—and as is proven in Section A.7—this lets us define a special system of coordinates "adapted" to the symmetry, in which $\xi = \frac{\partial}{\partial t}$. The Killing equation (4.32) then reduces to

$$\xi^t \partial_t g_{\mu\nu} = 0,$$

and the components of the metric are independent of our special time coordinate.

This notion of stationarity captures some of our intuition of what "not changing in time" means, but not quite all of it. Consider a uniform disk rotating at a constant angular velocity. Its geometry is independent of time, in the sense that any moment looks exactly like any other moment, but there is still clearly a sense in which the system is not at rest. In particular, under a time reversal, the direction of rotation reverses. To rule out such situations, we can add a requirement of invariance under reversal of our preferred time coordinate. A spacetime obeying this condition is not merely stationary, but "static."

General Relativity: A Concise Introduction. Steven Carlip © Steven Carlip 2019.
Published in 2019 by Oxford University Press. DOI: 10.1093/oso/9780198822158.001.0001

> **Box 10.1 Birkhoff's theorem**
>
> The derivation of the Schwarzschild geometry in this chapter assumes that the metric is static. As Birkhoff showed in 1923, this assumption is not really necessary; spherical symmetry is already strong enough. In fact, the Schwarzschild metric is the only spherically symmetric solution of the vacuum Einstein field equations; there are no additional time-dependent solutions that retain all of the symmetry. In particular, this implies that a spherically symmetric source cannot emit gravitational radiation, a result we have already seen in the weak field approximation.

In our adapted coordinates, it is evident that the metric components $g_{00}dt \otimes dt$ and $g_{ij}dx^i \otimes dx^j$ are time-reversal invariant, since the components are independent of t. The cross-terms $g_{i0}dx^i \otimes dt$, on the other hand, change sign under $t \to -t$, and must be excluded. The general static metric, in appropriate coordinates, thus takes the form

$$ds^2 = g_{00}(x^k)dt^2 + g_{ij}(x^k)dx^i dx^j . \tag{10.1}$$

10.2 Spherical symmetry

The Schwarzschild metric is also spherically symmetric. Formally, this means it admits three more Killing vectors, which act as the generators of rotations. We will again take a shortcut:

1. Fix the time t, and pick a point p. Its orbit under the rotation group—the set of points that can be reached from p by a rotation—will generically be a two-sphere. For spherical symmetry, the metric on this sphere must be the "round" metric of Section 2.1,

$$ds_2^2 = C(d\theta^2 + \sin^2\theta d\varphi^2) \text{ with } C \text{ independent of } \theta \text{ and } \varphi.$$

2. These spheres foliate space: almost every point lies on one such sphere, and two spheres never intersect. Pick a coordinate r to label the spheres; that is, let r be a coordinate that is constant on any single sphere but takes a different value on each different sphere.
3. On any given sphere, the metric components $(g_{r\theta}, g_{r\varphi})$ form the components of a two-dimensional vector, which picks out a preferred direction. But by spherical symmetry, no preferred direction can exist, so we must have $g_{r\theta} = g_{r\varphi} = 0$.
4. The component g_{00} cannot depend on θ of φ, since if it did, $(\partial_\theta g_{00}, \partial_\varphi g_{00})$ would again pick out a preferred direction. Similarly, g_{rr} cannot depend on θ or φ.

Combining these conditions, we obtain a metric

$$ds^2 = A(r)dt^2 - B(r)dr^2 - C(r)(d\theta^2 + \sin^2\theta d\varphi^2) . \tag{10.2}$$

There is still a remaining coordinate freedom: since r is just a label for spheres, we can trade r for any monotonic function of r. Standard coordinate choices include

- $B = 1$: These are "proper distance" coordinates: at constant (t, θ, φ), r is the proper radial distance.
- $C = r^2$: These are "areal" coordinates: at constant (t, r), the area of a sphere is $4\pi r^2$.
- $C = Br^2$: These are "isotropic" coordinates: at fixed t, the spatial metric is just C times the flat metric, so at any point, all three "Cartesian" directions are treated identically.
- $A = 1$: These are "Gaussian normal" or "synchronous" or "proper time" coordinates: at constant spatial position, t is the proper time.

While we can make any one of these choices, we cannot make more than one simultaneously. Each choice gives a different radial coordinate, with its own physical meaning, and the predictions of the theory will have to account for that meaning.

10.3 Schwarzschild again

Solving the vacuum Einstein field equations is easiest in areal coordinates, $C = r^2$. The computation is fairly routine, but it provides a good exercise in using the Cartan formalism of Section 5.9. We can read off an orthonormal basis one-form from the metric:

$$e^0 = \sqrt{A}\, dt, \quad e^1 = \sqrt{B}\, dr, \quad e^2 = r\, d\theta, \quad e^3 = r \sin\theta\, d\varphi, \qquad (10.3)$$

where the indices $\{0, 1, 2, 3\}$ label the basis vectors. The first Cartan structure equation (5.63) becomes

$$de^0 + \omega^0{}_\mu \wedge e^\mu = \frac{1}{2}\frac{A'}{\sqrt{A}} dr \wedge dt + \sqrt{B}\,\omega^0{}_1 \wedge dr + r\,\omega^0{}_2 \wedge d\theta + r\sin\theta\,\omega^0{}_3 \wedge d\varphi = 0,$$

$$de^1 + \omega^1{}_\mu \wedge e^\mu = \sqrt{A}\,\omega^1{}_0 \wedge dt + r\,\omega^1{}_2 \wedge d\theta + r\sin\theta\,\omega^1{}_3 \wedge d\varphi = 0,$$

$$de^2 + \omega^2{}_\mu \wedge e^\mu = dr \wedge d\theta + \sqrt{A}\,\omega^2{}_0 \wedge dt + \sqrt{B}\,\omega^2{}_1 \wedge dr + r\sin\theta\,\omega^2{}_3 \wedge d\varphi = 0,$$

$$de^3 + \omega^3{}_\mu \wedge e^\mu = \sin\theta\, dr \wedge d\varphi + r\cos\theta\, d\theta \wedge d\varphi$$
$$+ \sqrt{A}\,\omega^3{}_0 \wedge dt + \sqrt{B}\,\omega^3{}_1 \wedge dr + r\,\omega^3{}_2 \wedge d\theta = 0, \qquad (10.4)$$

where a prime denotes a derivative with respect to r. With a little practice, one can simply read off the connection one-form (keeping in mind that the wedge product of two one-forms is antisymmetric):

$$\omega^0{}_1 = \frac{1}{2}\frac{A'}{\sqrt{AB}}\, dt, \quad \omega^2{}_1 = \frac{1}{\sqrt{B}}\, d\theta, \quad \omega^3{}_1 = \frac{1}{\sqrt{B}} \sin\theta\, d\varphi, \quad \omega^3{}_2 = \cos\theta\, d\varphi \qquad (10.5)$$

with the remaining independent components vanishing.

The curvature two-form is then determined directly from the second Cartan structure equation (5.64):

> **Box 10.2 Interior solutions**
>
> This chapter focuses on the Schwarzschild metric as a solution of the vacuum field equations. To describe the gravitational field of a spherical body—a star or a planet, say—one must join this solution to an "interior solution" of the field equations (6.14) that accounts for the stress-energy tensor of the source.
>
> For a star, the stress-energy tensor may be approximated by that of a perfect fluid, eqn (7.8), with orthonormal basis components $T^0{}_0 = \rho(r)$, $T^i{}_j = -p(r)\delta^i_j$. The $G^0{}_0$ component of the field equations depends only on B and ρ, and gives
>
> $$B = \left(1 - \frac{2m(r)}{r}\right)^{-1} \quad \text{with } m(r) = 4\pi G \int_0^r \rho(r')r'^2 dr'. \qquad (10.6)$$
>
> The remaining field equations, supplemented by an equation of state, determine $A(r)$, $\rho(r)$, and $p(r)$. Details depend on the equation of state, but generically, the gravitational fields predicted by general relativity are stronger than their Newtonian counterparts, leading to new instabilities and the possibility of gravitational collapse.

$$\mathcal{R}^0{}_1 = d\omega^0{}_1 = -\frac{1}{2}\frac{1}{\sqrt{AB}}\left(\frac{A'}{\sqrt{AB}}\right)' e^0 \wedge e^1 = R^0{}_{101}\, e^0 \wedge e^1,$$

$$\mathcal{R}^0{}_2 = \omega^0{}_1 \wedge \omega^1{}_2 = -\frac{1}{2r}\frac{A'}{AB} e^0 \wedge e^2 = R^0{}_{202}\, e^0 \wedge e^2,$$

$$\mathcal{R}^0{}_3 = \omega^0{}_1 \wedge \omega^1{}_3 = -\frac{1}{2r}\frac{A'}{AB} e^0 \wedge e^3 = R^0{}_{303}\, e^0 \wedge e^3,$$

$$\mathcal{R}^1{}_2 = d\omega^1{}_2 + \omega^1{}_3 \wedge \omega^3{}_2 = \frac{1}{2r}\frac{B'}{B^2} e^1 \wedge e^2 = R^1{}_{212}\, e^1 \wedge e^2,$$

$$\mathcal{R}^1{}_3 = d\omega^1{}_3 + \omega^1{}_2 \wedge \omega^2{}_3 = \frac{1}{2r}\frac{B'}{B^2} e^1 \wedge e^3 = R^1{}_{313}\, e^1 \wedge e^3,$$

$$\mathcal{R}^2{}_3 = d\omega^2{}_3 + \omega^2{}_1 \wedge \omega^1{}_3 = \frac{1}{r^2}\left(1 - \frac{1}{B}\right) e^2 \wedge e^3 = R^2{}_{323}\, e^2 \wedge e^3, \qquad (10.7)$$

which immediately gives the components of the curvature tensor in our orthonormal basis. These contract to form a Ricci tensor

$$R_{00} = \frac{1}{2}\frac{1}{\sqrt{AB}}\left(\frac{A'}{\sqrt{AB}}\right)' + \frac{1}{r}\frac{A'}{AB},$$

$$R_{11} = -\frac{1}{2}\frac{1}{\sqrt{AB}}\left(\frac{A'}{\sqrt{AB}}\right)' + \frac{1}{r}\frac{B'}{B^2},$$

$$R_{22} = R_{33} = -\frac{1}{2r}\frac{A'}{AB} + \frac{1}{2r}\frac{B'}{B^2} + \frac{1}{r^2}\left(1 - \frac{1}{B}\right), \qquad (10.8)$$

with all other components identically zero.

The vacuum Einstein field equations require that $R_{ab} = 0$. Combining R_{00} and R_{11}, we see that
$$\frac{A'}{A} + \frac{B'}{B} = 0.$$
Hence AB is a constant, which we can set to 1 by rescaling A (that is, by rescaling the time coordinate). Then
$$R_{22} = 0 = \frac{1}{r}\frac{B'}{B^2} + \frac{1}{r^2}\left(1 - \frac{1}{B}\right) \Rightarrow A = \frac{1}{B} = 1 - \frac{2m}{r}, \qquad (10.9)$$
where m is an integration constant.

Inserting this back into the metric (10.2), we recover the Schwarzschild line element (3.1), now as an exact solution of the vacuum field equations. To give a physical interpretation to the integration constant m, we can look at the Newtonian limit (Section 2.5), evaluate the orbits (Chapter 3), or, as we shall see later, compute the value of the Hamiltonian (Section 12.6).

10.4 The event horizon

As we saw in Chapter 3, the Schwarzschild line element gives a very good description of the Solar System outside the Sun. If we try to extend the solution to small enough values of r, though, something goes wrong. At $r = 2m$, the component g_{tt} of the metric goes to zero, the component g_{rr} blows up, and the first integral (3.9) of the geodesic equation ceases to make sense. In the early days of general relativity, this was called the "Schwarzschild singularity."

But a "singularity" can have two origins. It may indicate a genuine breakdown of a physical theory, but it may instead mean nothing worse than a bad choice of coordinates. In ordinary polar coordinates, the component $g_{\theta\theta}$ of the metric goes to zero at the origin, and its inverse blows up, but there's nothing wrong with the geometry; the problem is just that the coordinate θ is not well defined there, leading to a "coordinate singularity."

For the Schwarzschild geometry, there are some clear hints that the "singularity" is not so bad. The components (10.7) of the curvature tensor in an orthonormal basis are perfectly well behaved at $r = 2m$, so the stresses on a freely falling observer, described by the geodesic deviation equation, remain finite. The Kretschmann scalar $R_{abcd}R^{abcd}$ doesn't blow up; neither do any other scalars built from the curvature tensor and its covariant derivatives. This is in marked contrast with the behavior at $r = 0$, where the curvature really does become infinite, an unambiguous sign of a genuine "curvature singularity."

To settle the question of whether $r = 2m$ is a true singularity or just a result of bad coordinates, we should search for a coordinate system in which the components of the metric behave better. The strategy will be to look for coordinates tied to real physical events, in this case the behavior of rays of light.

Consider an ingoing spherical shell of light. The motion of such a shell is described by a collection of ingoing null geodesics, each at a constant angular position, so at each (θ, φ) we have

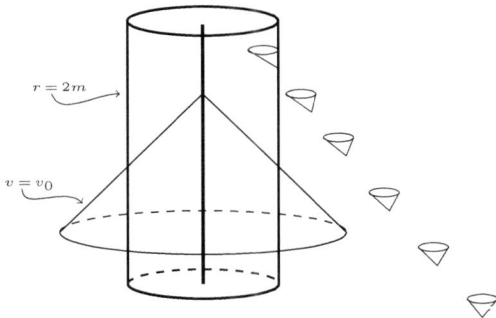

Fig. 10.1 Eddington-Finkelstein ingoing coordinates

$$ds^2 = 0 \Rightarrow dt = -\left(1 - \frac{2m}{r}\right)^{-1} dr \Rightarrow t = -r_* + v, \tag{10.10}$$

where the sign is fixed by the fact that the shell is ingoing, v is an integration constant, and the quantity r_*, the "tortoise coordinate," is defined as

$$r_* = r + 2m \ln\left|\frac{r}{2m} - 1\right|. \tag{10.11}$$

The integration constant v labels which shell of light we are looking at. But this also makes it a coordinate, a curved space analog of advanced time in electrodynamics. Instead of using t to label time, we can use v: we identify the time of an event by recording which ingoing shell of light is passing as the event occurs (see Fig. 10.1).

The new coordinates (v, r, θ, φ) are known as Eddington-Finkelstein advanced, or ingoing, coordinates. In this coordinate system, the Schwarzschild line element is

$$ds^2 = \left(1 - \frac{2m}{r}\right) dv^2 - 2dv dr - r^2(d\theta^2 + \sin^2\theta d\varphi^2), \tag{10.12}$$

as one can check by substituting $t = -r_* + v$ into the line element (3.1). The components are no longer "singular"; although g_{vv} goes to zero at $r = 2m$, the determinant and the components of the inverse metric remain finite. We can repeat the analysis with outgoing shells of light, using retarded time $u = t - r_*$. Eddington-Finkelstein retarded, or outgoing, coordinates (u, r, θ, φ) then give a line element

$$ds^2 = \left(1 - \frac{2m}{r}\right) du^2 + 2du dr - r^2(d\theta^2 + \sin^2\theta d\varphi^2), \tag{10.13}$$

which is again well behaved at $r = 2m$.

The nature of the surface $r = 2m$ is still a bit mysterious. We can learn more by reexamining radial null geodesics in Eddington-Finkelstein advanced coordinates. We now have

$$ds^2 = 0 \Rightarrow dv dr = \frac{1}{2}\left(1 - \frac{2m}{r}\right) dv^2. \tag{10.14}$$

One set of solutions is $v = v_0$, where v_0 is a constant. These are the ingoing shells we began with. The other is

$$\frac{dr}{dv} = \frac{1}{2}\left(1 - \frac{2m}{r}\right). \tag{10.15}$$

At large r, these are just ordinary outgoing light shells. But as r approaches $2m$, the slope changes—the light cones tilt inward, as shown in Fig. 10.1. In fact, the locus $r = 2m$ is itself a solution of (10.15), a shell of null geodesics.

To put it provocatively, the surface $r = 2m$ is expanding outward at the speed of light, but doing so without changing its area—something that is obviously impossible in flat spacetime, but that can occur if the curvature is large enough. Since nothing can move faster than light, nothing at $r < 2m$ can ever catch up to $r = 2m$ and escape. Thus $r = 2m$ is an event horizon, a "surface of no return."

For one more coordinate choice, we can use both u and v. More precisely, let

$$U = -e^{-\kappa u}, \quad V = e^{\kappa v}, \tag{10.16}$$

where $\kappa = 1/4m$ is called the surface gravity. In terms of the timelike Killing vector $\xi = \frac{\partial}{\partial t}$, κ may be given a geometrical form that extends to more complex black holes,

$$\kappa = \left(-\frac{1}{2}\nabla^a \xi^b \nabla_a \xi_b\right)^{1/2}\bigg|_{r=2m}. \tag{10.17}$$

The coordinates (U, V, θ, φ) are known as Kruskal-Szekeres coordinates, or sometimes just Kruskal coordinates. In these coordinates, the line element becomes

$$ds^2 = \frac{32m^3}{r}e^{-r/2m}dU\,dV - r^2(d\theta^2 + \sin^2\theta\, d\varphi^2), \tag{10.18}$$

where r is now a function of the coordinates U and V, defined implicitly by the equation

$$UV = -e^{r/2m}\left(\frac{r}{2m} - 1\right). \tag{10.19}$$

(This equation can be solved by using the Lambert W function, but the resulting expression is not very helpful.) It is now clear that nothing singular happens at $r = 2m$: a coordinate U or V goes to zero, but the metric is completely well-behaved.

10.5 An extended spacetime

Let us examine the Kruskal-Szekeres line element a little more carefully. By (10.19), the event horizon $r = 2m$ is located at $UV = 0$, that is, either $U = 0$ or $V = 0$. The curvature singularity $r = 0$ is at $UV = 1$, a double-branched hyperbola. Quite generally, any event at fixed (t, r) in Schwarzschild coordinates appears twice. We seem to have accidentally cloned our original spacetime. What we have discovered is an "extension" of the original manifold—technically, a larger manifold in which the Schwarzschild geometry can be isometrically embedded.

Figure 10.2 shows the structure of spacetime in Kruskal-Szekeres coordinates. Since U and V are null coordinates, their axes are drawn as 45° lines. Time flows from the

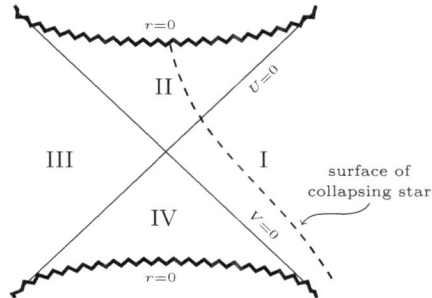

Fig. 10.2 The Kruskal-Szekeres extension

bottom to the top, while radial light rays move along 45° lines of constant U or V. Jagged lines are the past and future singularities. Each point on the diagram is really a two-sphere; the figure shows a cross-section of constant (θ, φ).

Regions I and III of this diagram are copies of the exterior of the black hole, $r > 2m$. Region II is the black hole interior. A physical object can pass into region II from the outside, falling through the horizon. Once inside, though, it can never again cross the $U = 0$ or $V = 0$ lines—that would require faster-than-light travel—but must inevitably reach the future singularity $r = 0$. Region IV is something new, a white hole, a time reversal of region II. Objects here can appear from the past singularity and escape to the exterior, but nothing from the exterior can enter.

As a practical matter, regions III and IV are irrelevant for real astrophysical black holes. The dashed line in Fig. 10.2 shows the surface of a star collapsing to form a black hole. To the left of that line is the interior of the star. That region is certainly not a vacuum, so the Schwarzschild solution is irrelevant there; for a correct description, the vacuum solution on the right would have to be joined to an "interior solution" as described in Box 10.2. Hence only the part of the diagram to the right of the dashed line—most of region I and part of region II—represents a physical solution of the field equations. On the other hand, it is widely speculated that a black hole formed by a quantum process might include the entire Kruskal-Szekeres spacetime.

10.6 Conformal structure and Penrose diagrams

Kruskal-Szekeres coordinates are very useful for visualizing the properties of black holes. But one important characteristic is still difficult to see: the behavior "at infinity." An event horizon, for instance, is defined by the condition that nothing inside the horizon can escape to infinity, and a more visual representation of this feature would be helpful. The way to achieve this is to rescale the metric, shrinking distances to bring "infinity" to a finite location. The resulting depiction is called a Penrose diagram, or Carter-Penrose diagram.

Start with the simplest example, two-dimensional Minkowski space. Introduce light cone coordinates $u = t - x$, $v = t + x$, so

$$ds^2 = du\, dv, \quad -\infty < u, v < \infty.$$

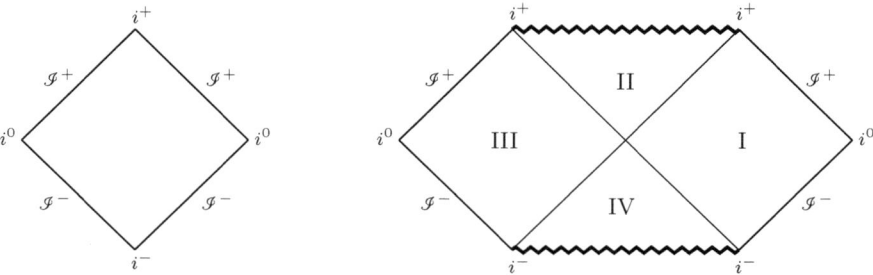

Fig. 10.3 Penrose diagrams for Minkowski space (left) and the Schwarzschild metric (right)

Next define new coordinates $\tilde{u}$ and $\tilde{v}$ by $u = \tan \tilde{u}$, $v = \tan \tilde{v}$, so

$$ds^2 = \sec^2 \tilde{u} \sec^2 \tilde{v} \, d\tilde{u} \, d\tilde{v}, \quad -\frac{\pi}{2} < \tilde{u}, \tilde{v} < \frac{\pi}{2}. \tag{10.20}$$

While $\tilde{u}$ and $\tilde{v}$ now have finite range, distances can still be infinite, since the prefactor $\sec^2 \tilde{u} \sec^2 \tilde{v}$ blows up at $\pm \pi/2$. Now, though, let us define a new spacetime by the transformation

$$g_{ab} \to \tilde{g}_{ab} = (\cos^2 \tilde{u} \cos^2 \tilde{v}) g_{ab} \;\Rightarrow\; d\tilde{s}^2 = d\tilde{u} \, d\tilde{v}, \quad -\frac{\pi}{2} < \tilde{u}, \tilde{v} < \frac{\pi}{2}. \tag{10.21}$$

"Infinity" in the old spacetime has now been rescaled to a finite distance in the new one, as shown in Fig. 10.3. In fact, "infinity" has been sorted into five regions: past timelike infinity i^-, the starting point of infinite timelike geodesics; future timelike infinity i^+, their endpoint; spacelike infinity i^0, the endpoint of spacelike geodesics; and past and future null infinity, $\mathscr{I}^-$ and $\mathscr{I}^+$ (pronounced "scri minus" and "scri plus"), the initial and final points of infinite null geodesics.

Transformations of the form $\tilde{g}_{ab} = e^{2\omega} g_{ab}$, called Weyl or conformal transformations, were briefly discussed in Section 5.8. They are *not* symmetries of the field equations; the metrics g_{ab} and $\tilde{g}_{ab}$ describe different geometries, and lead to different physics. But conformally related metrics have the same null geodesics and light cones. As long as we are only interested in the causal structure—which events are in the past or future of which other events—we are free to make such transformations.

The second diagram in Fig. 10.3 shows the Penrose diagram of the black hole, obtained from the Kruskal-Szekeres line element (10.18) in exactly the same way. The interior causal structure is essentially the same as that of Fig. 10.2, but now the structure at infinity is also made explicit. As before, null geodesics are represented by 45° lines, making it clear that light can escape to $\mathscr{I}^+$ from regions I, III, and IV but not from region II.

10.7 Properties of the horizon

The Schwarzschild solution is of interest in its own right: as we saw earlier, it is the starting point for many of the crucial tests of general relativity. But it is also a simple

Box 10.3 The laws of black hole mechanics

In 1973, Bardeen, Carter, and Hawking observed that stationary black holes obey four laws that are remarkably similar to the laws of thermodynamics [82]. For any black hole of mass M, angular momentum J, charge Q, horizon area A, and surface gravity κ (as defined in (10.17)),

0. The surface gravity is constant over the horizon;
1. For two stationary black holes that differ only by small variations in the parameters M, J, and Q,

$$\delta M = \frac{\kappa}{8\pi G}\delta A + \Omega_H \delta J + \Phi_H \delta Q,$$

where Ω_H and Φ_H are the angular velocity and electric potential at the horizon;

2. The area of the horizon never decreases;
3. It is impossible to reduce the surface gravity to zero by any physical process in a finite number of steps.

Bekenstein and Hawking subsequently showed that the resemblance to thermodynamics is not just coincidence: once quantum effects are included, a black hole really does radiate as a black body, with a temperature and entropy

$$kT = \frac{\hbar\kappa}{2\pi}, \quad S = \frac{A}{4\hbar G}.$$

The resulting field of black hole thermodynamics has offered crucial insights into quantum gravity.

---

example of a black hole, whose properties generalize to more complicated settings. Many features of the horizon $r = 2m$ extend to more complex black holes: the charged Reissner-Nordström solution, the rotating Kerr solution, the accreting Vaidya solution, higher dimensional black strings and black branes, and more elaborate interacting solutions that we know only as approximations. In particular:

- **The horizon is an event horizon.**

  The event horizon marks the point of no return, the boundary beyond which no signal can escape to infinity. From the Penrose diagram, we see that no light ray originating from region II can ever reach future null infinity $\mathscr{I}^+$.

  As one can see from Fig. 10.3, the past of $\mathscr{I}^+$—the region from which a causal signal can reach $\mathscr{I}^+$—consists of regions I, III, and IV. A general event horizon is defined as "the boundary of the past of future null infinity." For the Schwarzschild solution, that boundary is precisely $r = 2m$, the hypersurface that separates regions I, III, and IV from region II.

90 Black holes

- **The horizon is a marginally trapped surface.**

  Imagine a spherical lattice studded with flashbulbs that are timed to fire simultaneously in the rest frame of the lattice. The flash will produce two shells of light, one moving inward and one outward. In a flat spacetime, it is clear that the area of the inner shell will shrink, while the outer shell will grow.

  For a lattice in a Schwarzschild spacetime, the area of the inner shell will again always shrink. But from (10.15), the area $4\pi r^2$ of the outer shell will vary as

  $$\frac{dA}{dv} = 4\pi(r - 2m). \tag{10.22}$$

  If the flashbulbs fire at a location $r < 2m$, both the ingoing and outgoing shells of light will shrink. Since the lattice can't move faster than light, it will be trapped between the light shells, and will itself have to collapse.

  For a small region of any surface, one can define bundles of ingoing and outgoing null geodesics normal to the surface. The fractional rate of change of cross-sectional area of such a bundle is called its expansion. A surface for which both the ingoing and outgoing expansions are everywhere negative is called a trapped surface. The Schwarzschild horizon is marginally trapped: the ingoing expansion is negative, while the outgoing expansion is zero.

- **The horizon is a Killing horizon.**

  The "timelike Killing vector" $\xi = \frac{\partial}{\partial t}$ that characterizes a stationary spacetime is, indeed, timelike for the Schwarzschild solution as long as $r > 2m$. For $r < 2m$, on the other hand, the signs of g_{tt} and g_{rr} change, and ξ becomes spacelike. This is sometimes described, perhaps a bit confusingly, by the slogan, "Time and space trade places at the horizon."

  To make this more rigorous, we can write ξ in Kruskal-Szekeres coordinates, which remain good at the horizon:

  $$\xi = \frac{\partial}{\partial t} = \frac{\partial U}{\partial t}\frac{\partial}{\partial U} + \frac{\partial V}{\partial t}\frac{\partial}{\partial V} \;\Rightarrow\; \xi^U = -\kappa U,\; \xi^V = \kappa V, \tag{10.23}$$

  so

  $$g(\xi,\xi) = 2g_{UV}\xi^U\xi^V = -\frac{2m}{r}e^{-r/2m}UV. \tag{10.24}$$

  From (10.19), we see that ξ is indeed timelike in the exterior region, spacelike in the interior, and—most importantly here—null at $r = 2m$.

  The Killing vector has a further feature: it is normal to the horizon. Consider the horizon at $U = 0$. A general path on the horizon can be written in the form $(U, V, \theta, \varphi) = (0, V(s), \theta(s), \varphi(s))$, with a tangent vector

  $$w = \left(0, \frac{dV}{ds}, \frac{d\theta}{ds}, \frac{d\varphi}{ds}\right).$$

  Hence for any vector w tangent to the horizon,

$$g_{\mu\nu}\xi^\mu w^\nu = g_{UV}(\xi^U w^V + \xi^V w^U) = g_{UV}(-\kappa U)\frac{dV}{ds} = 0,\qquad(10.25)$$

since $U = 0$ at the horizon. The same holds for the horizon at $V = 0$; in both cases, the Killing vector is orthogonal to all tangent vectors, and thus normal to the horizon.

A Killing horizon is defined as a null hypersurface—that is, a hypersurface with a null normal—whose normal coincides with a Killing vector. This captures the intuition that a Killing vector changes from being timelike on one side of the horizon to spacelike on the other, generalizing the Schwarzschild result to more complicated settings.

For the Schwarzschild black hole, these three characteristics all describe the same horizon $r = 2m$. In more complicated situations, they may differ—a dynamical black hole may not have a Killing horizon, for instance—but their generalizations provide powerful tools. In particular, the discovery of the "four laws of black hole mechanics," described in Box 10.3, was one of the main spurs that led to the modern understanding of Hawking radiation and black hole thermodynamics.

Further reading

A treatment of spherical symmetry starting with the Killing equation may be found in Chapter 6 of [35], which also contains a discussion of the invariant definition of a static spacetime. The Schwarzschild metric was first derived in [15], initially in a somewhat different coordinate system. Birkhoff's theorem was published in [83]; a proof can be found in Section 5.2 of [17].

Interior solutions are discussed in more detail in part V of [11], as well as Chapter 10 of [10]. Birkhoff's theorem does not apply when the stress-energy tensor is nonzero, and explicit solutions describing gravitational collapse can be constructed. The first such solution, for pressureless matter, was found by Oppenheimer and Snyder in 1939 [84]. Subsequent work has required a mixture of analytic and numerical methods; see, for example, [85].

Eddington-Finkelstein coordinates originated in [86] and [87]. Kruskal-Szekeres coordinates originated in [88] and [89]. Carter-Penrose diagrams have their roots in [90] and [91].

An excellent introduction to the mathematical techniques necessary to understand black holes is Poisson's book, *A Relativist's Toolkit* [42]. Chapter 12 of [35], especially Section 12.5, may also be useful. Frolov and Zelnikov's book, *Black Hole Physics*, offers a good overview [92]. For a broad summary and review of black hole thermodynamics, see [93], and for a wide-ranging list of resources regarding black holes, see [94].

This chapter has not described the observational evidence for the existence of real astrophysical black holes. For a brief introduction, see [95]. The first observation of gravitational waves from a black hole merger is reported in the LIGO discovery paper [81], which includes an explanation of why the source was almost certainly a black hole binary.

11
Cosmology

In the preceding chapter, we managed to solve the Einstein field equations by imposing symmetries strong enough to reduce the ten partial differential equations to two ordinary differential equations. We now turn to a second setting in which a similar simplification occurs. We know from observation that at the scales larger than about 100 Mpc, our Universe is highly uniform. This implies a new set of symmetries, which are again strong enough to let us find exact solutions of the field equations.

The search for cosmological solutions dates back to the earliest days of general relativity. A few crucial observations—the expansion of the Universe and, later, the cosmic microwave background and the abundance of light elements—made early tests possible. More recently, cosmology has become far more data-rich, allowing for the construction and testing of much more elaborate and detailed models. This chapter will only briefly touch on these recent and rapidly changing developments. Rather, the aim is again to provide tools for further exploration.

11.1 Homogeneity and isotropy

The observed uniformity of the large scale Universe implies two distinct symmetries, homogeneity and isotropy.

- A spacelike hypersurface Σ_t (see Section 2.2) at a fixed time t is said to be homogeneous if any small region of Σ_t is indistinguishable from any other.
- A spacetime is said to be isotropic around a point p if it looks the same in every direction, that is, if there is no "preferred direction" at p.

Homogeneity and isotropy are not the same. The electric field of a point charge is isotropic around the charge, but it varies with distance, so it is not homogeneous. The electric field of an infinite charged plate is homogeneous, but it picks out a direction, so it is not isotropic. Note that each symmetry requires a choice: a time coordinate to identify the hypersurfaces of homogeneity, a rest frame for an observer at the point p to eliminate directional Doppler shifts that would spoil isotropy.

These symmetries strongly restrict the metric. Pick coordinates (t, x^i) such that the hypersurfaces of constant t are homogeneous. The line element must then take the form

$$ds^2 = N^2(t)dt^2 + g_{ij}dx^i dx^j . \tag{11.1}$$

By homogeneity, N cannot depend on the spatial coordinates; otherwise regions with different values would be distinguishable. By isotropy, no cross terms g_{0i} can occur; otherwise these would form the components of a vector on Σ_t and pick out a preferred direction.

General Relativity: A Concise Introduction. Steven Carlip © Steven Carlip 2019.
Published in 2019 by Oxford University Press. DOI: 10.1093/oso/9780198822158.001.0001

Homogeneity and isotropy

Next pick a fixed time t and consider the curvature tensor $^{(3)}R_{ij}{}^{kl}$ of the three-dimensional manifold Σ_t. The superscript (3) indicates that this is the intrinsic curvature of a hypersurface itself, built solely out of g_{ij} and its spatial derivatives. Each pair of indices $[ij], [kl]$ in $^{(3)}R_{ij}{}^{kl}$ is antisymmetric, and thus has only three independent values. We can isolate these by defining

$$^{(3)}\tilde{R}^m{}_n = \tilde{\epsilon}^{mij}\tilde{\epsilon}_{nkl}\,^{(3)}R_{ij}{}^{kl}, \tag{11.2}$$

where $\tilde{\epsilon}$ is the Levi-Civita symbol (4.19) and the indices m, n run from 1 to 3.

The components $^{(3)}\tilde{R}^m{}_n$ form a symmetric 3×3 matrix, with three eigenvectors and three eigenvalues. If the eigenvalues were not all equal, the eigenvector corresponding to the highest (or lowest) eigenvalue would pick out a preferred direction, violating isotropy. To avoid this, we must demand

$$^{(3)}\tilde{R}^m{}_n = \lambda \delta^m_n \quad \Rightarrow \quad ^{(3)}R_{ij}{}^{kl} = \frac{\lambda}{4}\left(\delta_i^k \delta_j^l - \delta_j^k \delta_i^l\right), \tag{11.3}$$

where the Bianchi identity (5.50) forces λ to be constant on Σ_t.

Spaces of this type are called "spaces of constant curvature," and have been studied extensively by mathematicians. Up to an overall scale, the metric can take three forms:

$$d\sigma_3^2 = \begin{cases} d\psi^2 + \sin^2\psi\,(d\theta^2 + \sin^2\theta\,d\varphi^2) & \text{three-sphere } S^3 \\ d\psi^2 + \psi^2\,(d\theta^2 + \sin^2\theta\,d\varphi^2) & \text{flat space } \mathbb{R}^3 \\ d\psi^2 + \sinh^2\psi\,(d\theta^2 + \sin^2\theta\,d\varphi^2) & \text{hyperbolic three-space } \mathbb{H}^3 \end{cases}, \tag{11.4}$$

or, in different coordinates,

$$d\sigma_3^2 = \frac{dr^2}{1 - kr^2} + r^2\,(d\theta^2 + \sin^2\theta\,d\varphi^2) \quad \begin{cases} k = 1 & \text{three-sphere } S^3 \\ k = 0 & \text{flat space } \mathbb{R}^3 \\ k = -1 & \text{hyperbolic three-space } \mathbb{H}^3 \end{cases}. \tag{11.5}$$

The constant curvature condition determines the spatial metric only up to a scaling; the general metric takes the form $a^2 d\sigma_3^2$. By homogeneity, a must be spatially constant, but it can depend on time. Putting the pieces together, we finally obtain a line element

$$ds^2 = N^2(t)dt^2 - a^2(t)d\sigma_3^2. \tag{11.6}$$

A bit of freedom still remains in the choice of time coordinate. The two most common optionsenormous are

- $N = 1$: These are "proper time" or "cosmological time" coordinates.
- $N = a$: These are "conformal time" coordinates. In these coordinates, time is usually denoted η.

The spatial coordinates (ψ, θ, φ) or (r, θ, φ) are called comoving coordinates. It is easy to check that a trajectory along which these coordinates remain constant is a geodesic. An observer on such a trajectory—a "comoving observer"—is locally at rest,

94 Cosmology

Box 11.1 The topology of the Universe

For the metrics (11.5) to be defined globally, the spatial topology cannot be completely arbitrary. The $k = 1$ metric describes a three-sphere, while the $k = 0$ and $k = -1$ metrics are defined on topologically trivial ("open") spaces. But as discussed in Section A.3, more complicated spatial topologies can be made by cutting out sections of these simple spaces and "gluing together" the edges. There is, in fact, a sense in which most three-manifold topologies can be obtained from a space with $k = -1$.

If space has a nontrivial topology, we should be able to look around a closed "circumference" and see the same object in two different directions. In practice, this is not so easy: besides having to look very far, we would typically see the "same" object after very different light travel times in the two directions, and therefore at two very different ages. The best bet may be to look for correlations in fluctuations of the cosmic microwave background [96]. So far, there is no sign that our Universe has a nontrivial large scale topology, but the possibility has not been ruled out, and observational searches are continuing.

---

to the extent this phrase has meaning. Such an observer sees an isotropic Universe, with no preferred direction, and cannot be described as having a velocity in any direction.

The coefficient $a(t)$—sometimes denoted $R(t)$—is the scale factor, or expansion factor. Imagine two radially separated comoving observers, at $(\psi_1, \theta, \varphi)$ and $(\psi_0, \theta, \varphi)$, where the subscript 0 conventionally means "now" or "us." Their proper distance is $\Delta s = a(t)(\psi_1 - \psi_0)$, which changes in time in proportion to the scale factor. Similarly, the volume of a comoving region is proportional to a^3. We thus have a peculiar situation in which two observers can each be locally at rest with respect to the the Universe, but nevertheless moving relative to each other.

We will also need the stress-energy tensor, which is again highly constrained by symmetry. By homogeneity, $T^0{}_0 = \rho(t)$, and by isotropy, $T^i{}_0 = 0$. By isotropy, the space-space components $T^i{}_j$ must have equal eigenvalues, and by homogeneity these can depend only on time, so $T^i{}_j = -p(t)\delta^i_j$. We are led back to the perfect fluid stress-energy tensor of Section 7.3, with the added restriction that ρ and p must be spatially constant.

11.2 The FLRW metric

Given the metric (11.6), the curvature tensor and Ricci tensor are easy to compute. In cosmological time $N = 1$, the Einstein tensor has the simple form

$$G^0{}_0 = \frac{3\dot{a}^2}{a^2} + \frac{3k}{a^2},$$

$$G^i{}_j = \left(\frac{2\ddot{a}}{a} + \frac{\dot{a}^2}{a^2} + \frac{k}{a^2}\right)\delta^i_j, \qquad (11.7)$$

where a dot indicates a time derivative. The field equations thus become

$$\frac{\dot{a}^2}{a^2} + \frac{k}{a^2} = \frac{8\pi G}{3}\rho, \qquad (11.8)$$

$$\frac{2\ddot{a}}{a} + \frac{\dot{a}^2}{a^2} + \frac{k}{a^2} = -8\pi G p. \qquad (11.9)$$

These are known as the Friedmann equations. Their solution is called a Friedmann-Lemaître-Robertson-Walker (FLRW) metric, or sometimes a Friedmann-Robertson-Walker or Robertson-Walker or Friedmann-Lemaître metric. The logarithmic derivative of the scale factor,

$$H = \frac{\dot{a}}{a}, \qquad (11.10)$$

is the Hubble constant, the fractional rate of expansion of the Universe. H is, of course, not usually constant in time, but at any fixed moment it is constant in space.

Differentiating (11.8) and combining with (11.9), we obtain a conservation law

$$\dot{\rho} + 3(\rho + p)\frac{\dot{a}}{a} = 0. \qquad (11.11)$$

For matter with an equation of state $p = w\rho$ with constant w, this implies

$$\rho \sim a^{-3(1+w)}. \qquad (11.12)$$

For noninteracting matter ($w = 0$), which cosmologists often call "dust," this is ordinary conservation: the density is inversely proportional to the volume. For radiation ($w = 1/3$), there is an additional fall-off factor: as the Universe expands, radiation is also red-shifted, reducing its energy by an extra factor of $1/a$. A cosmological constant ($w = -1$) is, indeed, constant. Note that as the Universe expands, species of matter with smaller values of w are diluted more slowly, and become more dominant. Conversely, in the very early Universe, species of matter with larger values of w dominate.

Inserting (11.8) directly into (11.9), we find an "acceleration" equation,

$$\frac{\ddot{a}}{a} = -\frac{4\pi G}{3}(\rho + 3p). \qquad (11.13)$$

As long as $\rho + 3p$ is positive, the presence of matter causes the expansion of the Universe to decelerate, as expected from the fact that gravity is attractive. But the effective energy density responsible for the acceleration is not just ρ, but $\rho + 3p$. The observed accelerating rate of expansion implies the existence of some type of "dark energy"—perhaps a cosmological constant—with $\rho + 3p < 0$.

We can also run this argument backwards. As we trace the Universe back to earlier times, its contraction will accelerate, and a will go to zero, giving an initial singularity. This is not just an accident of symmetry; the singularity theorems of Hawking and Penrose show that with a fairly minimal set of assumptions, the existence of an initial singularity is unavoidable. Of course, this claim should not be taken too literally—when curvatures become large near a putative singularity, we expect quantum effects

> **Box 11.2 de Sitter and anti-de Sitter space**
>
> Cosmological solutions of the Einstein field equations are especially simple if the only "matter" is a cosmological constant. The solution with $\Lambda > 0$ is de Sitter space, or dS. It can be described by an FLRW metric with $k = 1$ and
>
> $$a = \ell \cosh \frac{t}{\ell} \quad \text{with } \Lambda = \frac{3}{\ell^2}.$$
>
> The solution with $\Lambda < 0$ is anti-de Sitter space, or AdS. It can be described by an FLRW metric with $k = -1$ and
>
> $$a = \ell \sin \frac{t}{\ell} \quad \text{with } \Lambda = -\frac{3}{\ell^2}.$$
>
> The de Sitter solution may be directly relevant to our Universe. If the Universe has a positive cosmological constant—a plausible explanation for the observed accelerated expansion—then it is almost certainly asymptotically de Sitter in the distant future. The anti-de Sitter solution does not seem to describe our Universe, but it has attracted wide interest because of its role in quantum gravity and string theory.

to make classical general relativity untrustworthy—but it points to important open questions about initial conditions.

For one more consequence of the Friedmann equations, let us define a dimensionless density parameter Ω_i for each species of matter (baryons, radiation, cold dark matter, and so on),

$$\Omega_i = \frac{8\pi G}{3H^2} \rho_i, \quad \Omega_{\text{tot}} = \sum_i \Omega_i. \tag{11.14}$$

Then we have

$$\frac{k}{a^2} = \Omega_{\text{tot}} - 1. \tag{11.15}$$

Thus Ω_{tot} measures the deviation from spatial flatness, with $k = 1$ for $\Omega_{\text{tot}} > 1$, $k = 0$ for $\Omega_{\text{tot}} = 1$, $k = -1$ for $\Omega_{\text{tot}} < 1$. We observe $\Omega_{\text{tot}} \approx 1$: the Universe is "nearly flat," with a density near the "critical density" $\rho_c = 3H^2/8\pi G$.

11.3 Observational implications

To solve the Friedmann equations, we need one more piece of information, the equation of state of the matter content of the Universe. This cannot come from general relativity, but must be determined observationally. The "standard model" of cosmology, consisting of ordinary baryonic matter ($w = 0$), cold dark matter ($w = 0$), radiation ($w = \frac{1}{3}$), and "dark energy" ($w \approx -1$), fits observations well. But even without a detailed census of matter, some predictions can be extracted from basic physics and the general form of the equations:

- **The Universe is expanding.**

  More precisely, the Universe is not static. Equations (11.8)–(11.9) have a solution with $\dot{a}=0$, the Einstein static Universe, but it requires a very precise tuning of (negative) pressure to balance density, and is unstable against small perturbations. Given an explicit description of the matter content of the Universe, we can solve the Friedmann equations to work out the details of the expansion history. Conversely, by measuring the expansion history, we can extract a good deal of information about the matter content.

- **Light from distant stars is red-shifted.**

  The proof is not very different from the derivation of red shift in Section 3.6. In the coordinates (11.4), radial light rays satisfy $ds^2 = 0 = dt^2 - a^2 d\psi^2$. Consider a source at ψ_1 and an observer at ψ_0. (As before, the subscript 0 denotes "us.") As in Section 3.6, suppose a first pulse of light is emitted at t_1 and received at t_0, and a second pulse is emitted at $t_1 + \Delta t_1$ and received at $t_0 + \Delta t_0$. Then

  $$\int_{\psi_1}^{\psi_0} d\psi = \int_{t_1}^{t_0} \frac{dt}{a} = \int_{t_1+\Delta t_1}^{t_0+\Delta t_0} \frac{dt}{a} \Rightarrow \int_{t_1}^{t_1+\Delta t_1} \frac{dt}{a} = \int_{t_0}^{t_0+\Delta t_0} \frac{dt}{a}. \quad (11.16)$$

  For small Δt, this means $\Delta t_1 / a(t_1) = \Delta t_0 / a(t_0)$. Converting from periods to frequencies, we obtain a red shift

  $$z = \frac{\omega_1 - \omega_0}{\omega_0} = \frac{a(t_0) - a(t_1)}{a(t_1)}. \quad (11.17)$$

  For fairly nearby galaxies, this gives $z \approx H(t_0 - t_1)$, Hubble's law.

- **The Universe was once very hot and dense.**

  This is a direct consequence of the expansion. The very early Universe was dominated by the matter with the highest value of w, radiation. By standard thermodynamics, the energy density of radiation varies with temperature as $\rho \sim T^4$, so from (11.12) with $w = 1/3$, we have $T \sim 1/a$. This has several implications for observation:

  - Light elements were produced in the very early Universe. At early enough times, the density and temperature were high enough for nuclear fusion to produce an assortment of elements (^{4}He, ^{2}H, ^{3}He, ^{7}Li). As the Universe expanded and the temperature fell, this "primordial nucleosynthesis" cut off, leaving remnants with abundances that can be calculated from nuclear physics.
  - Electromagnetic radiation from the early Universe is still present, and has a thermal spectrum. After primordial nucleosynthesis, the Universe was filled with a plasma of electrons, protons, and some light nuclei, all at thermal equilibrium with radiation. As the temperature dropped further, a critical value was reached at which the electrons were able to fairly abruptly combine with protons and other nuclei to form atoms. Plasmas are opaque to light, but gases are nearly transparent, so after this time—called "last scattering" or "recombination"—the radiation was able to move freely.

98 *Cosmology*

The subsequent expansion of the Universe has red-shifted this radiation, but has preserved the thermal form of its spectrum. The black body spectrum depends on frequency and temperature in the combination ω/kT, and as we have just seen, ω and T have the same dependence on the scale factor, so expansion leaves the shape of the spectrum unchanged.

This remnant radiation is the cosmic microwave background, or CMB. Such an analysis cannot by itself tell us the temperature of the CMB now, because it cannot tell us when "now" is. But we can check for consistency with a chain of arguments:

1. Nucleosynthesis depends on both temperature and density. From the observed abundance of ^{4}He, we can determine both quantities at the era of primordial nucleosynthesis. These predictions can be checked against observed abundances of other light elements.
2. We then use the Friedmann equations to evolve forward to the time of last scattering. We know the temperature at last scattering—it's the temperature at which hydrogen ionizes—so the theory gives us the density.
3. We evolve the rest of the way, to "now." We determine the present density from observation; the theory then gives us the CMB temperature. The result agrees with what we see.

Note that without this explanation, it is extremely hard to understand why the CMB has so nearly perfect a black body spectrum. While one can imagine other sources of background radiation, it is virtually impossible to obtain a black body spectrum except from a system that has been in thermal equilibrium.

- **Primordial perturbations determine both the deviation of the CMB from perfect isotropy and the growth of structure.**

While the Universe is homogeneous at large scales, at small scales it certainly isn't. It's filled with "structure"—clusters of galaxies, galaxies, stars, planets, us. And while the CMB is very nearly thermal and isotropic, it's not quite—it has a particular, very accurately measured spectrum of fluctuations.

The variations of the CMB are almost certainly a consequence of random fluctuations in the extremely early Universe, plausibly quantum fluctuations of the vacuum blown up by inflation (see Section 11.4). These same fluctuations should also have led to density perturbations that acted as seeds of structure formation, attracting surrounding matter and growing in size and mass. Although we have not discussed it here, it is not too hard to compute the behavior of perturbations of an FLRW metric, using methods similar to those of Chapter 8. A major area of research in both theoretical and observational cosmology is the effort to check for consistency between the fluctuations of the CMB and the growth of structure.

11.4 Inflation

While the picture of cosmology we have developed is very attractive, it leaves several important questions unanswered. One of the most puzzling is the "horizon problem." A simple way to see this problem is to consider the Penrose diagram for an FLRW metric

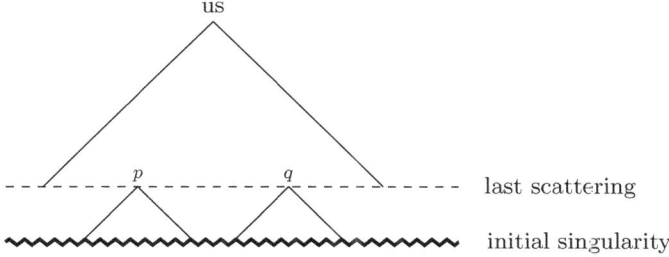

Fig. 11.1 Conformal diagram for an FLRW metric

(see Section 10.6). This is most easily obtained by writing the metric in conformal time η rather than cosmological time t. The FLRW metric is then conformal to

$$d\tilde{s}^2 = d\eta^2 - d\sigma_3^2 \tag{11.18}$$

with a Penrose diagram shown in Fig. 11.1. The diagram shows the initial singularity, or whatever replaces it in quantum gravity; the surface of last scattering, the origin of the cosmic microwave background; and present-day observers, "us."

Consider cosmic microwave background radiation originating at points p and q in the diagram. We can observe the CMB from both points, since both are within our past light cone. But p and q have never been in causal contact: no signal from anywhere in the Universe ever had time to reach both. It is puzzling, then, that the temperature and fluctuation spectrum of the CMB coming from these two points should agree so precisely.

Of course, whether this is a problem or not depends on details of the diagram. If the spacing between the initial singularity and the surface of last scattering were large enough, there would be more time for p and q to communicate, and the puzzle might disappear. For our current models of the matter content of the Universe, though, regions just a few degrees apart in the sky should have never been in causal contact.

Inflationary cosmology addresses this puzzle by positing a period of nearly exponential expansion in the extremely early Universe. In the conformal diagram 11.1, a long enough period of inflation has the effect of pushing the initial singularity down far enough to allow points like p and q to equilibrate. The same mechanism can also help with the "flatness problem," the fact that the observed density parameter Ω_{tot}, discussed in Section 11.2, is so close to one. In a Universe with $\rho + 3p > 0$, it is not hard to check that $\Omega_{\text{tot}} = 1$ is a repulsive fixed point of the dynamics—that is, even if Ω_{tot} starts close to one at some early time, the dynamics will push it away. During inflation, though, $\rho + 3p < 0$, and the dynamics will rapidly drive any initial density parameter towards one. This, of course, is not a complete answer—the present value still depends on the initial conditions, which we don't know—but it makes the observation that $\Omega_{\text{tot}} \approx 1$ seem more reasonable.

Perhaps most intriguingly, inflation provides a natural source for the observed fluctuations of the CMB. In quantum mechanics, even a field in its vacuum state is subject to vacuum fluctuations. These normally occur on a scale much too small to be relevant to cosmology. But an early inflationary period can blow these fluctuations up exponentially, producing a spectrum that agrees remarkably well with the complicated observed pattern of variations of the CMB.

While inflationary models are quite attractive, though, the mechanism for inflation remains a mystery. One possibility is the presence of a new scalar field, dubbed the "inflaton." If a scalar field φ is nearly constant, the potential term in the stress-energy tensor (7.11) acts approximately as a cosmological constant, leading to de Sitter-like expansion (Box 11.2). By adjusting the potential, and through that the dynamics, one can easily create models with an early period of inflation. Other possibilities are also available; for instance, an extra term in the Einstein-Hilbert action proportional to a square of the curvature scalar can produce an inflationary cosmology. Observations have now reached the point that they can start to constrain models, but there is not yet a compelling argument for any particular mechanism.

Further reading

For a much more comprehensive description of relativistic cosmology, see, for example, Weinberg's *Cosmology* [97] or Dodelson's *Modern Cosmology* [98]. Carroll's general relativity textbook [17] also has a strong emphasis on cosmological applications.

A mathematical treatment of spaces of constant curvature can be found in [99]. To understand possible spatial topologies, a good starting point is [34]; see also [100].

The FLRW metric comes from work by Friedmann [101], Lemaître [102], Robertson [103], and Walker [104]. See Chapter 6 of Longair's *The Cosmic Century* for some of the history [105]. The proof of the "singularity theorems" mentioned in Section 11.2 requires global geometric techniques, which are discussed briefly in Section 13.2 below; see that chapter for references. More details about de Sitter and anti-de Sitter space may be found in [106, 107].

Cosmology, like gravitational radiation, is a rapidly evolving field. Again, the journal *Living Reviews in Relativity*, http://www.springer.com/us/livingreviews, is a good source for reasonably current review articles.

12
The Hamiltonian formalism

So far, our approach to general relativity has been based on the Lagrangian formalism. Elsewhere in physics, it is often useful to switch to a Hamiltonian formalism, both for issues of principle—the Hamiltonian approach is the starting point for canonical quantization, for instance—and as a practical means of simplifying certain calculations. This chapter will introduce the Hamiltonian formulation of general relativity.

12.1 The ADM metric

In standard Hamiltonian mechanics, time plays a special role: the definition of generalized positions and momenta requires a choice of time coordinate, and Hamilton's equations of motion are equations for time derivatives. We now want to apply this formalism to general relativity, where there is no preferred time coordinate. To do so, we will have to temporarily break the symmetry between space and time by choosing a "time-slicing"—a foliation of spacetime into spacelike hypersurfaces Σ_t labeled by some time coordinate—and then check at the end that the physics is independent of that choice.

Given a time-slicing, the construction of the Arnowitt-Deser-Misner (ADM) metric is essentially an application of the Lorentzian version of Pythagoras' theorem, as shown in Fig. 12.1. To find the proper time between points (t, x^i) and $(t + dt, x^i + dx^i)$:

1. Choose two nearby slices Σ_t and Σ_{t+dt}, each with a positive definite spatial metric. Call these metrics $q_{ij}(t)$ and $q_{ij}(t + dt)$.
2. Pick an initial point (t, x^i) on Σ_t and move orthogonally to Σ_{t+dt}. The elapsed proper time will not be simply dt—after all, t is an arbitrary coordinate—but will take the form Ndt for some function N. N is called the lapse function.
3. This motion will end at a point on Σ_{t+dt} whose spatial coordinates are not necessarily just $\{x^i\}$—these are again arbitrary coordinates—but may be shifted to nearby values $x^i - N^i dt$. N^i is called the shift vector.
4. On Σ_{t+dt}, perform the additional displacement to $x^i + dx^i$.

Combining the spatial interval on Σ_{t+dt} with the temporal interval, we end up with the ADM line element,

$$ds^2 = N^2 dt^2 - q_{ij}(dx^i + N^i dt)(dx^j + N^j dt). \tag{12.1}$$

This expression has the same number of independent functions as the usual spacetime metric—six from the spatial metric, three from the shift vector, one from the lapse function—but they are now split into a clear "(3 + 1)-dimensional" description. We

102 The Hamiltonian formalism

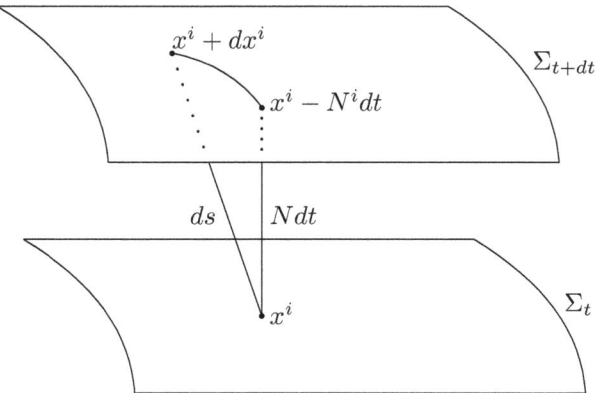

Fig. 12.1 The geometry of the ADM metric

now raise and lower spatial indices with q_{ij} and its inverse q^{kl} (see Box 12.1). A straightforward calculation from (12.1) then shows that the four-metric has an inverse

$$g^{00} = \frac{1}{N^2}, \quad g^{0i} = -\frac{N^i}{N^2}, \quad g^{ij} = -q^{ij} + \frac{N^i N^j}{N^2}, \qquad (12.2)$$

and its determinant has a square root

$$\sqrt{|g|} = N\sqrt{q}, \qquad (12.3)$$

where q is the determinant of the matrix of components q_{ij}.

12.2 Extrinsic curvature

From the Hamiltonian viewpoint, the q_{ij} are generalized positions. The generalized momenta are normally time derivatives of the positions. But $\dot{q}_{ij}$ won't quite do: the time derivatives of the components of a tensor don't form a tensor. To find the canonical momenta, we need one more geometric detour.

Up to now, the curvature we have discussed has been "intrinsic curvature," a quantity completely determined by the intrinsic geometry of a manifold. We have made no assumptions about how, or whether, that manifold sits inside any higher dimensional space. But if a manifold *is* embedded in a higher dimensional space, there is another kind of curvature as well. A two-dimensional cylinder, for instance, may have no intrinsic curvature—you can roll a sheet of paper into a cylinder without stretching or otherwise distorting its intrinsic geometry—but it does have "extrinsic curvature."

Just as intrinsic curvature is defined in terms of covariant derivatives of tangent vectors, extrinsic curvature is defined in terms of covariant derivatives of the normal vector. A plane in flat three-space has no extrinsic curvature; a cylinder does, because its normal changes direction.

Box 12.1 Conventions: part VII

In the standard four-dimensional view of general relativity, the spacetime manifold is the fundamental geometric structure. We raise and lower indices with the metric $g_{\mu\nu}$, take covariant derivatives ∇ with the connection compatible with $g_{\mu\nu}$, and so on. In the Hamiltonian formalism, it is often convenient to treat the spatial slice Σ as the fundamental geometric structure. So we define q^{ij} as the matrix inverse of q_{kl}, raise and lower spatial indices with q_{ij}, take covariant derivatives $^{(3)}\nabla$ with the connection compatible with q_{ij}, and so on. Often when there is an ambiguity, a superscript $^{(3)}T$ or $^{(4)}T$ to the left of an object labels the dimension. But these conventions are not completely standard; some authors retain four-dimensional notation in the Hamiltonian formalism. The overall sign of the extrinsic curvature tensor (12.4) can also vary, and the relative sign of the two terms depends on whether the hypersurface is timelike ($n_a n^a = 1$ with our choice of metric signature) or spacelike ($n_a n^a = -1$).

Let n_a be the unit normal to a spacelike hypersurface Σ. The extrinsic curvature tensor is defined to be

$$K_{ab} = \nabla_a n_b - n_a n^c \nabla_c n_b \,. \tag{12.4}$$

While K_{ab} is a four-dimensional tensor, it has no components normal to Σ: $n^b K_{ab} = 0$ because n^b is a unit vector, and $n^a K_{ab} = 0$ because of the last term in (12.4). It is not obvious, but it may be checked that K_{ab} is symmetric.

For the metric in the ADM form, the unit normal to surface of constant t is

$$n_\mu = (N, 0, 0, 0) \,. \tag{12.5}$$

A simple calculation then gives

$$K_{ij} = -\frac{1}{2N} \left(\partial_0 q_{ij} - {}^{(3)}\nabla_i N_j - {}^{(3)}\nabla_j N_i \right) \,. \tag{12.6}$$

(There are also components K_{i0} and K_{00}, but these will cancel out in later expressions.) The trace, $K = q^{ij} K_{ij}$, is often called the "mean curvature." We thus have a quantity that contains the time derivative of the spatial metric but that also, by construction, is tensorial.

We now come to a key result. The Gauss-Codazzi equation of classical differential geometry relates the intrinsic and extrinsic curvature of an embedded manifold to the curvature of the manifold in which it is embedded. In its original form, applied to surfaces in flat three-dimensional space, this was Gauss's "theorema egregium," or "remarkable theorem." For us, it says that, with our conventions,

$$^{(4)}R = -\,{}^{(3)}R - K_{ij}K^{ij} + K^2 + 2\,{}^{(4)}\nabla_a v^a$$
$$\text{with } v^a = n^b({}^{(4)}\nabla_b n^a) - n^a({}^{(4)}\nabla_b n^b) \,, \tag{12.7}$$

where, as above, $K = q^{ij} K_{ij}$, and $^{(3)}R$ is the curvature scalar of the spatial metric q_{ij}.

12.3 The Hamiltonian action

We can now substitute the curvature scalar (12.7) into the Einstein-Hilbert action (6.6), setting the cosmological constant to zero for simplicity. The last term is a total derivative (see Section 5.5), and will not contribute to the bulk action. Thus

$$I_{\text{grav}} = -\frac{1}{2\kappa^2} \int dt \int_\Sigma d^3x \sqrt{q} N \left(-^{(3)}R - K_{ij}K^{ij} + K^2\right) \tag{12.8}$$

with $\kappa^2 = 8\pi G$. As in standard Hamiltonian mechanics, the momentum conjugate to the generalized position q_{ij} is

$$\pi^{ij} = \frac{\partial \mathscr{L}}{\partial(\partial_0 q_{ij})}, \tag{12.9}$$

where the Lagrangian density $\mathscr{L}$ is the integrand of the action. But the action depends on $\partial_0 q_{ij}$ only through the extrinsic curvature, so

$$\pi^{ij} = \frac{1}{2\kappa^2} \sqrt{q} N \frac{\partial}{\partial(\partial_0 q_{ij})} \left(K_{ij}K^{ij} - K^2\right) = -\frac{1}{2\kappa^2} \sqrt{q}(K^{ij} - q^{ij}K). \tag{12.10}$$

The Hamiltonian is thus

$$H = \int_\Sigma d^3x \left(\pi^{ij}\partial_0 q_{ij} - \mathscr{L}\right) = \int_\Sigma d^3x \left(N\mathscr{H} + N_i\mathscr{H}^i\right), \tag{12.11}$$

where, after a bit of calculation,

$$\mathscr{H}^i = -2\,^{(3)}\nabla_j \pi^{ij}, \tag{12.12}$$

$$\mathscr{H} = \frac{2\kappa^2}{\sqrt{q}} \left(\pi^{ij}\pi_{ij} - \frac{1}{2}\pi^2\right) - \frac{1}{2\kappa^2}\sqrt{q}\,^{(3)}R, \tag{12.13}$$

with $\pi = q_{ij}\pi^{ij}$. The quantity $\mathscr{H}^i$ is known as the momentum constraint or the diffeomorphism constraint, while $\mathscr{H}$ is the Hamiltonian constraint. The Hamiltonian form of the action is

$$I_{\text{grav}} = \int dt \int_\Sigma d^3x \left(\pi^{ij}\partial_0 q_{ij} - N\mathscr{H} - N_i\mathscr{H}^i\right). \tag{12.14}$$

We can now treat general relativity as an ordinary Hamiltonian system, albeit with some slightly unusual features to be discussed below. We define Poisson brackets on a fixed time slice,

$$\{q_{ij}(\mathbf{x}), \pi^{kl}(\mathbf{x}')\} = \frac{1}{2}\left(\delta_i^k \delta_j^l + \delta_j^k \delta_i^l\right) \delta^{(3)}(\mathbf{x} - \mathbf{x}'), \tag{12.15}$$

and write the standard set of Hamilton's equations

$$\partial_0 q_{ij} = \{q_{ij}, H\}, \quad \partial_0 \pi^{ij} = \{\pi^{ij}, H\}, \tag{12.16}$$

which can be shown to be equivalent to the Einstein field equations. Note that in the process, we have reduced a set of second order partial differential equations to first order, a very useful feature for numerical implementation.

12.4 Constraints

For the most part, the theory defined in the preceding section is a conventional Hamiltonian theory. It has a few peculiar features, though. To begin with, the lapse and shift have no canonically conjugate momenta. This can be traced back to the fact that their time derivatives never appear in the action. In fact, N and N_i are Lagrange multipliers: when varied in the action (12.14), they lead to equations

$$\mathcal{H} = 0, \quad \mathcal{H}^i = 0. \tag{12.17}$$

These equations are constraints. They contain no time derivatives of the canonical variables, but instead constrain the allowed data (q_{ij}, π^{ij}) on an initial hypersurface. In fact, general relativity is a "completely constrained" system, whose Hamiltonian consists entirely of constraints. When the equations of motion are satisfied, the Hamiltonian thus has the value zero, up to possible boundary terms that will be considered in Section 12.6. This does not mean the Poisson brackets (12.16) are zero—they involve not just the Hamiltonian, but its functional derivatives—but it does have important implications for the interpretation of the theory.

To see this, observe that the constraints also serve a second function: they generate the "gauge invariances" of the theory. More precisely, let $\xi^\perp$ and $\hat{\xi}^i$ be arbitrary scalar and vector fields on Σ. Define the smeared constraints

$$\mathcal{H}[\xi^\perp] = \int_\Sigma d^3x\, \xi^\perp \mathcal{H}, \quad \hat{\mathcal{H}}[\hat{\xi}] = \int_\Sigma d^3x\, \hat{\xi}^i \mathcal{H}_i, \tag{12.18}$$

and consider their Poisson brackets with an arbitrary functional $F[q, \pi]$,

$$\delta_{(\xi^\perp, \hat{\xi})} F = \left\{ F, \mathcal{H}[\xi^\perp] + \hat{\mathcal{H}}[\hat{\xi}] \right\}. \tag{12.19}$$

Such a transformation is called a "surface deformation." With a bit of work, it is possible to show that on shell—that is, when the equations of motion are satisfied— a surface deformation parametrized by $\xi^\perp$ and $\hat{\xi}^i$ is equivalent to a diffeomorphism parametrized by a vector field ξ^μ, with

$$\xi^\perp = N\xi^0, \quad \hat{\xi}^i = \xi^i + N^i \xi^0. \tag{12.20}$$

In other words, the constraints are essentially the canonical generators of diffeomorphisms. This makes intuitive sense: if the constraints are zero on shell, there ought to be some sense in which their Poisson brackets are also zero, and the action of a diffeomorphism is indeed "zero" in the sense that it does not change the physics.

This feature is not unique to gravity. A system of constraints is called first class if it is closed under Poisson brackets. In his analysis of constrained Hamiltonian systems, Dirac showed that very generally, first class constraints generate gauge invariances. Box 12.2 illustrates this property for the simpler case of electromagnetism.

12.5 Degrees of freedom

The Hamiltonian formalism makes it easy to count the degrees of freedom of general relativity. We start with twelve dynamical degrees of freedom (q_{ij}, π^{ij}). The constraints

> **Box 12.2 Constraints and gauge transformations in electromagnetism**
>
> Consider an electromagnetic field in flat spacetime, with an action (7.15):
>
> $$I_{\text{EM}} = -\int dt \int_\Sigma d^3x \left(\frac{1}{2}F_{0i}F^{0i} + \frac{1}{4}F_{ij}F^{ij}\right). \tag{12.21}$$
>
> As in Section 12.3, the canonical momentum is $\pi^i = \frac{\partial \mathscr{L}}{\partial(\partial_0 A_i)} = -F^{0i}$, which can be recognized as the electric field. The Hamiltonian is
>
> $$H = \int_\Sigma d^3x \left(\pi^i \partial_0 A_i - \mathscr{L}\right) = \int_\Sigma d^3x \left(-\frac{1}{2}F_{0i}F^{0i} + \frac{1}{4}F_{ij}F^{ij} + \pi^i \partial_i A_0\right). \tag{12.22}$$
>
> The first two terms are the energy (7.17). The third, after integration by parts, is a constraint. Like the lapse and shift, the component A_0 occurs with no time derivatives, and acts as a Lagrangian multiplier for the constraint $\partial_i \pi^i = 0$. This constraint may be recognized as Gauss's law, which is indeed a restriction on initial data rather than a dynamical equation.
>
> The smeared constraint, the electromagnetic equivalent of (12.18), is
>
> $$\mathcal{C}[\Lambda] = \int_\Sigma d^3x \, (\partial_i \Lambda) \pi^i. \tag{12.23}$$
>
> Its Poisson brackets with the canonical fields are
>
> $$\{A_i, \mathcal{C}[\Lambda]\} = \partial_i \Lambda, \quad \{\pi^i, \mathcal{C}[\Lambda]\} = 0, \tag{12.24}$$
>
> which are indeed the gauge transformations of the potential and the electric field. The extension to non-Abelian fields is a bit more complicated, but not too hard.

give four relations, leaving eight independent degrees of freedom. But four of these can be fixed by a choice of coordinates, leaving four "free" phase space degrees of freedom, that is, two configuration space degrees of freedom and their time derivatives. This count was already apparent in Section 9.2, where we saw that gravitational waves have only two independent polarizations.

Of course, being able to count degrees of freedom is not the same as being able to explicitly characterize them. In some cases, we can go further, using a method known as the York decomposition. We must restrict ourselves to spacetimes that admit a "constant mean curvature" time-slicing, that is, a time-slicing for which K is constant on each slice and for which $K = -t$ is a good time coordinate. Not every spacetime admits such a slicing, but a large class do, providing one of the few known instances in which a purely geometrical quantity can act as a "clock." In cosmology, this choice amounts to using the local rate of expansion of the Universe as a time coordinate.

It can be shown that the "free data" needed to specify a solution of the constraints on such a slice consist of a pair $(\bar{q}_{ij}, p^{ij})$, where $\bar{q}_{ij}$ is a Yamabe metric, that is, a metric of constant scalar curvature, and p^{ij} is a transverse traceless tensor density,

$$^{(3)}R[\bar{q}] = 0 \text{ or } \pm 1, \quad \bar{\nabla}_i p^{ij} = 0, \quad \bar{q}_{ij} p^{ij} = 0. \tag{12.25}$$

As described in Section A.8, the physical three-metric and momentum can be reconstructed from the free data and a conformal factor ϕ, which is determined by a remaining constraint, the Lichnerowicz-York equation. This equation is mathematically well behaved and has a unique solution, but it cannot usually be solved in closed form. The description of the initial data is thus not completely explicit, but the simplification is enormous. This method has been crucial for constructing initial data for numerical relativity.

12.6 Boundary terms

So far, we have written the smeared Hamiltonian and momentum constraints as spatial volume integrals, integrating by parts when necessary. It is often the case, though, that these expressions require additional boundary terms. The fundamental requirement for any region with a boundary, including an asymptotic boundary at infinity, is that the Poisson brackets (12.19) of the smeared constraints be well-defined. This, in turn, requires that the variations of the smeared constraints be "integrable":

$$\delta \mathcal{H}[\xi] = \int_\Sigma d^3x \left(\frac{\delta \mathcal{H}[\xi]}{\delta q_{ij}} \delta q_{ij} + \frac{\delta \mathcal{H}[\xi]}{\delta \pi^{ij}} \delta \pi^{ij} \right) \tag{12.26}$$

with no surface terms coming from integration by parts. If surface terms *do* appear, they must be cancelled by suitable boundary terms added to the constraints. Without such boundary terms, the Poisson brackets (12.19) are valid only up to delta functions at the boundary, and the standard Hamiltonian formalism breaks down.

Consider the simplest case, an asymptotically flat spacetime. In asymptotically Cartesian coordinates, we expect

$$q_{ij} \sim \delta_{ij} + h_{ij} \quad \text{with } h_{ij} = \mathcal{O}(1/r), \tag{12.27}$$

and we will restrict to smearings $\xi^\perp \sim 1$, that is, to transformations that are asymptotically time translations. Let us integrate over a region Σ defined asymptotically by the condition $r < r_0$, whose boundary $\partial \Sigma$ is approximately a sphere of radius r_0; we will take the limit $r_0 \to \infty$ at the end.

The dangerous term in the variation (12.26) comes from the spatial curvature $^{(3)}R$ in the Hamiltonian constraint. We have already analyzed the variation of curvature scalar in Section 6.2; copying from there, we expect boundary terms from the total derivatives

$$\delta \mathcal{H}[\xi^\perp] = \cdots - \frac{1}{2\kappa^2} \int_\Sigma d^3x \sqrt{q} \, \xi^\perp \, ^{(3)}\nabla_k \left[q^{ij} \delta(^{(3)}\Gamma^k_{ij}) - q^{ki} \delta(^{(3)}\Gamma^l_{il}) \right]$$

$$= \cdots - \frac{1}{2\kappa^2} \int_\Sigma d^3x \sqrt{q} \, \xi^\perp \, ^{(3)}\nabla_k \left[q^{ij} q^{kl} \left(^{(3)}\nabla_i \delta q_{jl} - ^{(3)}\nabla_l \delta q_{ji} \right) \right]$$

$$= \cdots - \frac{1}{2\kappa^2} \int_{\partial \Sigma} d^2x \sqrt{q} \, \xi^\perp \, q^{ij} n^l \left(^{(3)}\nabla_i \delta q_{jl} - ^{(3)}\nabla_l \delta q_{ij} \right), \tag{12.28}$$

where n^l is the unit normal to the surface $r = r_0$. By (12.27), the asymptotic behavior of this expression is

$$\delta \mathcal{H}[\xi^\perp] \sim \cdots - \frac{1}{2\kappa^2} \int_{\partial \Sigma} d^2x \, \delta^{ij} n^l \left(\partial_i \delta q_{jl} - \partial_l \delta q_{ji} \right) \tag{12.29}$$

(taking into account that $\xi^\perp \sim 1$). This does not vanish as $r_0 \to \infty$: the integrand goes as $1/r_0^2$, but we are integrating over a sphere of area $\sim r_0^2$. But the variation can be canceled by adding a boundary term to the Hamiltonian constraint,

$$\mathcal{H}_\infty = \frac{1}{2\kappa^2} \int_\infty d^2x \, \delta^{ij} n^l \left(\partial_i q_{jl} - \partial_l q_{ji} \right), \tag{12.30}$$

where the integral is over the sphere at spatial infinity. This is the ADM Hamiltonian.

We can evaluate $\mathcal{H}_\infty$ for an asymptotically Schwarzschild metric, for which

$$q_{ij} \sim \left(1 + \frac{2GM}{r}\right) \delta_{ij}, \quad \partial_k q_{ij} \sim -\frac{2GM}{r^2} \delta_{ij} \partial_k r = -\frac{2GM}{r^2} \delta_{ij} n_k. \tag{12.31}$$

Inserting this into (12.30), we obtain

$$\mathcal{H}_\infty = \frac{1}{2\kappa^2} \int_\infty d^2x \left(\frac{4GM}{r^2}\right) = \frac{16\pi GM}{2\kappa^2} = M. \tag{12.32}$$

For more general asymptotically flat spacetimes, $\mathcal{H}_\infty$ is called the ADM mass.

The same analysis can be applied to regions with other kinds of boundaries. At future null infinity $\mathscr{I}^+$ (see Section 10.6), the boundary term gives the Bondi mass, which provides an invariant measure of energy carried off by gravitational waves. For finite regions, the boundary term can be used to define "quasilocal energy," the best we can do to localize gravitational energy (see Box 7.1). There have also been interesting attempts to find boundary terms at the horizon of a black hole, which might give us better insight into universal properties of horizon symmetries.

Further reading

The history of the Hamiltonian formalism of general relativity is more complex than is often appreciated—see, for example, [108]—but the final formulation is due to Arnowitt, Deser, and Misner [109]. Dirac's analysis of constraints and their relationship to gauge symmetries appeared in [110]. For a comprehensive review, see Henneaux and Teitelboim's *Quantization of Gauge Systems* [111]. A nice discussion of the surface deformations and their geometrical significance can be found in [112].

The York decomposition of Section 12.5 was introduced in [113]. For a more recent summary of its status, see [114]. A broad introduction to the (3+1)-dimensional formalism appears in Gourgoulhon's book, *3+1 Formalism in General Relativity* [115], as well as Chapter 21 of [11].

The importance of integrability of the constraints and its relevance for boundary terms was developed by Regge and Teitelboim [116]. For some applications to boundaries at finite distances, see [62].

13
Next steps

General relativity is a broad and vibrant field of research. This book has only touched on the basics; it should be treated as a launching point for further exploration. This final chapter will offer a brief introduction to a few areas worthy of deeper study.

13.1 Solutions and approximations

The Einstein field equations are ten coupled nonlinear partial differential equations, each with hundreds of terms, and we are unlikely to find the general solution without radical new insights. The exact solutions we have seen here, the Schwarzschild solution and the FLRW cosmologies, were obtained by imposing symmetries that drastically reduced the number of degrees of freedom. There are many other possible symmetries, which can be classified in various ways, most notably by algebraic properties of the Weyl tensor (the Petrov classification) or by their Lie algebras. These, in turn, have led to hundreds of exact solutions.

A few of these are clearly physically important: the Kerr solution for rotating black holes, other axially symmetric solutions describing stars with higher multipole moments, interior Schwarzschild solutions for the internal structure of stars. Others may have physical relevance: slightly inhomogeneous and nonisotropic cosmologies, for instance, and the "Mixmaster" solution for spacetimes near an initial cosmological singularity. Yet others offer "existence theorems" (exact gravitational wave solutions) or examples of interesting behavior that may generalize (formation of event horizons, new types of singularities, asymptotic behaviors).

To go further, one must resort to approximation methods. The weak field, slow motion post-Newtonian approximation discussed in Chapter 8 can be understood as an expansion in powers of v/c, with $GM/rc^2 \sim v^2/c^2$. The nth post-Newtonian approximation—"nPN" for short—is the expansion to order $(v/c)^{2n}$. Chapter 8 gave the 1PN order; the state of the art (as of 2018) is 4PN. In some circumstances, other approximations may be useful. Weak fields can be expanded around backgrounds other than flat spacetime; the post-Minkowskian expansion assumes weak fields but not necessarily slow motion; and for systems such as a small star orbiting a supermassive black hole, an expansion in powers of the mass ratio can be instructive.

The most significant progress in recent years has occurred in numerical relativity. While perturbation theory can be a powerful method for investigating weak fields, the only reliable way to extract predictions in the strong field regime is to use numerical techniques. The effort to put general relativity on a computer dates back to 1964, but it was only in 2005 that Frans Pretorius, quickly followed by others, developed

General Relativity: A Concise Introduction. Steven Carlip © Steven Carlip 2019.
Published in 2019 by Oxford University Press. DOI: 10.1093/oso/9780198822158.001.0001

a stable, long term simulation of binary black hole orbits and mergers. Since then, the field has grown explosively. Most numerical work has focused on binary systems of compact objects (black holes or neutron stars), where it has been instrumental in predicting gravitational waveforms. But numerical methods have also been used to study gravitational collapse and supernovae, neutron star equations of state, critical phenomena, the behavior of the gravitational field near a singularity, and a variety of other issues in astrophysics and cosmology.

13.2 Mathematical relativity

Many important issues can be formulated as questions about the mathematical structure of the Einstein field equations. Some topics in mathematical relativity include:

- **The initial value problem:**

  We saw in Chapter 12 that general relativity is a constrained system: initial data cannot be chosen arbitrarily, but are subject to the Hamiltonian and momentum constraints. As described in Section 12.5, these constraints are under good control for a large class of geometries. But for others, those that do not admit a slice of constant mean curvature, the constraints are much more poorly understood.

  Give smooth initial data obeying the constraints, Yvonne Choquet-Bruhat showed in the 1950s that the Einstein field equations are well-posed, that is, that they have a unique local solution. Questions about how far to the future such solutions extend, how smooth the initial data need to be, and whether solutions are stable against small perturbations of initial data are active topics of research.

- **Singularity theorems:**

  The exact solutions we have seen here, the Schwarzschild solution and the FLRW cosmologies, have singularities. For years it was an open question whether these were artifacts of the strong symmetry assumptions, but in the 1960s Hawking and Penrose showed that under quite general circumstances singularities were inevitable. The proofs required the creation of a new toolkit of mathematical techniques to investigate the global structure of spacetimes, which have now become an integral part of general relativity. The singularity theorems do not tell us much about the detailed nature of the singularities, though; this is again an active area of research.

  Proofs of singularity theorems typically require "energy conditions," restrictions on the stress-energy tensor that enforce some sort of positivity of energy. The weak energy condition, for example, is the requirement that $T_{ab}u^a u^b \geq 0$ for any timelike vector u. The classical energy conditions can typically be violated in quantum field theory, and there is a continuing effort to find the appropriate quantum extensions.

- **Cosmic censorship:**

  The singularity of the Schwarzschild metric is hidden behind a horizon, and cannot be seen from the outside. The weak cosmic censorship conjecture is the hypothesis that this feature is generic, that is, that singularities that form from generic initial data are invisible from future null infinity.

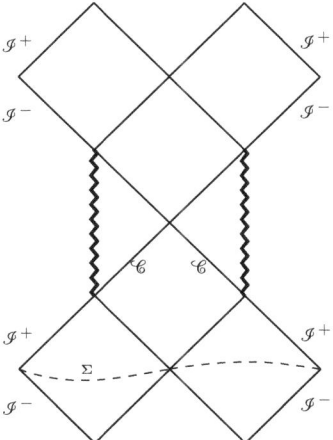

Fig. 13.1 A portion of the Penrose diagrams for a charged black hole

Charged and rotating black holes have a rather different kind of singularity, a timelike singularity, as illustrated in the Penrose diagram of Fig. 13.1. Such singularities lead to a breakdown of predictability: information can enter the spacetime from the singularity, so initial data no longer uniquely determine future evolution. Initial data on the surface Σ in Fig. 13.1, for instance, can be evolved up to $\mathscr{C}$, the "Cauchy horizon," but beyond that the evolution becomes indeterminate. Cauchy horizons—boundaries of predictability—can also occur in nonsingular spacetimes, for instance in regions with closed timelike curves. The strong cosmic censorship conjecture is the hypothesis that Cauchy horizons do not occur generically, that is, that for generic initial data singularities will appear (perhaps behind horizons) to stop the evolution before a Cauchy horizon can form.

The terminology is historical, and a bit misleading: "strong" cosmic censorship is not stronger than "weak" cosmic censorship, and neither conjecture implies the other. Various versions of censorship have been shown to hold in restricted models, but the general question is open. Other types of "censorship" have also been proposed; the topological censorship theorem, for instance, shows that local topological features of isolated systems are hidden behind horizons.

- **Asymptotic behavior:**

  The Penrose diagrams of Fig. 10.3 show that even flat Minkowski space has a more elaborate structure at infinity than one might expect. This sort of structure leads to an important set of research topics: what possible asymptotic structures exist, how they form from initial data, how the metric and curvature tensor fall off near infinity. Of particular interest are asymptotic symmetries and corresponding conserved quantities, which generalize the ADM mass of Section 12.6. For d-dimensional asymptotically anti-de Sitter space, the asymptotic symmetry group is the $(d-1)$-dimensional conformal group, a fact that plays a central role in the AdS/CFT correspondence between gravity and conformal quantum field theory.

For asymptotically flat space, the symmetry at null infinity is the Bondi-Metzner-Sachs (BMS) group, which may be important in understanding quantities such as scattering amplitudes.

13.3 Alternative models

So far, no experiment or observation has found a fundamental flaw in general relativity. There are, however, puzzles. Especially in cosmology, the combination of theory and observation seems to require additional, poorly understood sources of gravity: "dark matter" and an even more mysterious "dark energy." It is natural to ask whether we are really seeing signs of modified gravity. Even without this impetus, one of the best ways to test a theory like general relativity is to compare it to alternative theories.

Section 6.4 briefly discussed possible geometrical generalization of the Einstein-Hilbert action: the inclusion of torsion, nonmetricity, and higher powers of curvature. Another possibility is to introduce nongeometrical fields that either mimic gravity or affect the way gravity couples to matter. An early example is Brans-Dicke theory, a scalar-tensor theory with an action

$$I = -\int_M \left(\phi R - \omega \frac{g^{ab}\nabla_a\phi\nabla_b\phi}{\phi} \right) \sqrt{|g|}\, d^4x + I_{\text{matter}}[g, \phi, \psi], \qquad (13.1)$$

where ω is a coupling constant and ϕ is a scalar field that acts as a varying gravitational constant. No experiment can completely rule out such models, but we can constrain possible couplings. For Brans-Dicke theory, for instance, current observations require ω to be at least 10^4 (general relativity is recovered when $\omega \to \infty$.) A more general form of scalar-tensor theory, Horndeski theory, may lead to interesting phenomenology. More elaborate generalizations can include vectors as well as scalars. Bimetric and multimetric theories are often inconsistent, having negative energy excitations, but a special class of consistent theories can lead to "massive gravity."

The possibilities are enormous, and considerable research has gone into sorting out candidates and exploring their implications for observation. For example, many models predict different speeds for gravitational and electromagnetic waves, so the simultaneous observation of gravitational waves and light from merging neutron stars has severely constrained certain parameters. Further limits come from a variety of Solar System and cosmological observations, but some models remain viable, and perhaps phenomenologically interesting.

13.4 Quantum gravity

As early as 1916, Einstein suggested that it might be necessary to combine general relativity with the emerging field of quantum mechanics. Quantum theory as we now know it did not yet exist—the Schrödinger equation and the Heisenberg operator formulation were still a decade away—but it was clear that something "quantum" was needed to stabilize atoms, which would otherwise collapse as they radiated away energy as electromagnetic radiation. Einstein pointed out that if this were true for electromagnetic radiation, it should be true for gravitational radiation as well.

While this is not quite the rationale we would use today, almost all physicists expect that a fully consistent coupling of gravity to quantum matter will ultimately require a quantum theory of gravity. The relevant scales are tiny, though. The Planck length, where quantum gravitational effects are expected to become important, is

$$\ell_p = \sqrt{\frac{\hbar G}{c^3}} \approx 1.6 \times 10^{-35} \text{ m}, \quad (13.2)$$

twenty orders of magnitude smaller than the size of a proton. We thus have little experimental guidance beyond the classical limit. As a problem of fundamental physics, though, quantum gravity seems unavoidable.

The decades-long search for such a theory has taught us a great deal about fundamental physics, but the final goal remains elusive. The problem is partly technical: we would certainly be further along if we knew the general solution to the Einstein field equations. But the difficulty is also rooted in the very different starting points of quantum mechanics and general relativity. To normalize wave functions in a quantum theory, we must require probabilities to sum to one "at a fixed time." To define local operators, we must know what it means to talk about a field "at a fixed location." To impose causality, we must require that operators commute if they are "spacelike separated." All of this requires a fixed spacetime background, very different from the dynamical spacetime of general relativity. If spacetime itself is quantized, such basic notions as "fixed time," "fixed location," and "spacelike" become unclear.

For instance, in ordinary canonical quantization, one writes a time-dependent Schrödinger equation in terms of a Hamiltonian. But as we saw in Section 12.4, the Hamiltonian in general relativity is a constraint, which vanishes for physical configurations. We can still try to follow the standard procedure, writing momenta as derivatives,

$$\pi^{ij} \to \frac{\hbar}{i} \frac{\delta}{\delta q_{ij}},$$

and turning the Hamiltonian constraint (12.13) into an operator equation. This yields the Wheeler-DeWitt equation,

$$\hat{\mathcal{H}} \Psi[q] = -\hbar \left(16\pi \ell_p^2 G_{ijkl} \frac{\delta}{\delta q_{ij}} \frac{\delta}{\delta q_{kl}} + \frac{1}{16\pi \ell_p^2} \sqrt{q}\,^{(3)}R \right) \Psi[q] = 0, \quad (13.3)$$

where ℓ_p is the Planck length (13.2) and $G_{ijkl} = \frac{1}{2}\frac{1}{\sqrt{q}}(q_{ik}q_{jl} + q_{il}q_{jk} - q_{ij}q_{kl})$ is the DeWitt metric on the space of (spatial) metrics. There are technical problems with this formulation—the product of functional derivatives is poorly defined, and we don't know how to make the space of functions $\Psi[q]$ into a Hilbert space with a sensible inner product—but something like it is likely to survive in any quantum theory of gravity. More problematic, though, is the physical interpretation. Our would-be Schrödinger equation contains no time derivatives, and it is not at all obvious how to extract anything like time evolution. This is an instance of the "problem of time" in quantum gravity, which is part of a more general "problem of observables": it can be shown that a diffeomorphism-invariant quantum theory of gravity can have no observables that are local in time or space.

While we do not yet have a complete, consistent quantum theory of gravity, we *do* have a number of interesting research programs. These include:

- **String theory:**

    String theory describes the metric, as well as the other elementary fields of particle physics, as excitations of one-dimensional strings, potentially unifying gravity with the other fundamental interactions. In asymptotically anti-de Sitter spacetimes (Box 11.2), string theory suggests a duality between quantum gravity in the interior and a nongravitational conformal field theory at the boundary at infinity, the AdS/CFT correspondence, which may perhaps be used to define what we mean by quantum gravity. More recent results have hinted at intriguing connections between spacetime locality and quantum entanglement.

- **Loop quantum gravity:**

    Loop quantum gravity is a canonical quantization program in which the basic configuration space variable is taken to be a generalized connection rather than a metric. Such a connection is naturally described by its holonomies around loops (see Section A.6). More general states are described by spin networks, graphs with edges labeled by group representations that are coupled at vertices. These are acted on by geometric operators such as area and volume. Spin networks propagating in spacetime form "spin foams," which may be useful in a path integral setting.

- **Discrete approaches:**

    In causal dynamical triangulations, the Feynman path integral over spacetime metrics is approximated as a sum over simplicial complexes, curved spacetimes built from flat building blocks in the same way a geodesic dome is built from flat triangles. An even more primitive approach, causal set theory, describes spacetime as simply a collection of discrete points, with no information beyond their causal relationships.

- **Quantum field theory:**

    The naive application of quantum field theoretic methods to general relativity fails. General relativity is nonrenormalizable: a consistent description requires an effective action with infinitely many terms, each with its own coupling constant. The asymptotic safety program hypothesizes that these constants might all be determined by a small number of parameters at high energy, at an ultraviolet fixed point of the renormalization group flow. If this is the case, it might allow us to treat quantum gravity as a more or less conventional quantum field theory.

Beyond these (and other) broad research programs, there are a variety of other ideas for exploring aspects of quantum gravity. These include the study of quantum black holes, the investigation of simpler models such as quantum gravity in two or three spacetime dimensions, and the search for "universal" characteristics of quantum gravity such as a possible minimum length or the proposed "dimensional reduction" to two-dimensional behavior near the Planck scale.

13.5 Experimental gravity

For much of its early history, the observational evidence for general relativity was quite weak. Until about 1960, apart from tests of the equivalence principle, the only real confirmations were the advance of Mercury's perihelion and the deflection of light, described in Chapter 3. These were joined in the early 1960s by the gravitational red shift and Shapiro time delay, to constitute the "four classical tests" of general relativity. Proposals existed for further tests—searches for frame-dragging and for gravitational waves, for instance—but these seemed out of reach.

This situation has changed dramatically. Experimental gravitation is now an extraordinarily active and wide-ranging field. The biggest impetus has come from the decades-long effort to directly detect gravitational waves, which finally succeeded in 2015. The laser interferometric detectors responsible for these observations, capable of measuring changes of distance of 10^{-18} m, have been called the most sensitive instruments ever built. Measurements of signals coming from mergers of stellar mass black holes and neutron stars have not only provided direct tests of general relativity in the strong field regime, but have opened up a new era in astronomy, in which we will be able to study events that would otherwise be completely invisible. Space-based laser interferometers and pulsar timing arrays will extend the reach, potentially allowing the detection of merging supermassive black holes throughout the Universe. At the same time, general relativistic predictions ranging from the speed and polarization of gravitational waves to the uniqueness of black hole configurations will be subject to increasingly sensitive tests.

But while the detection of gravitational radiation is certainly the most dramatic advance in experimental gravitation, it is far from the only one. In the laboratory, increasingly accurate experiments have allowed us to test the principle of equivalence for individual components of energy: nuclear binding energy, electrostatic and magnetostatic energy, energy carried by weak neutral currents, kinetic energy of electrons in atoms, and so on. Other experiments are searching for violations of the inverse square law at very short distances, an effect predicted by many alternative models of gravity. Gravitational time dilation tests, using atomic clocks and neutron and atom interferometry, look for variations that might depend on distance, time, velocity, or the physical processes involved in the clocks, any one of which could indicate a problem with the theory. There are even proposals to probe quantum gravity, by looking at gravitational fields of quantum superpositions of masses at different locations. By eliminating many sources of noise, space-based versions of these experiments may provide orders of magnitude higher precision.

At larger scales, Lunar laser ranging can now measure the distance to the Moon with millimeter precision, allowing tests of the Nordtvedt effect (see Box 8.1) and searches for variations of Newton's constant. Lunar laser ranging also allows us to observe geodetic precession, a precession of the spin of a gyroscope—the "gyroscope" in this case being the Earth-Moon system—under parallel transport. Geodetic precession, also known as de Sitter precession, has also been seen by the satellite experiment Gravity Probe B, which simultaneously measured "frame-dragging" (see Section 8.3). Proposed interplanetary laser ranging—for instance to Mars's moon Phobos—could drastically increase precision. Short of that, Mercury orbiters will allow much better observations of the perihelion advance, while the Gaia mission's measurement of the positions of more than a billion stars should permit an extraordinarily accurate test of the deflection of light by the Sun.

Yet another set of tests come from measurements of binary stellar systems. Binary pulsars are especially useful, since a pulsar acts as an extremely accurate orbiting clock. Gravitational radiation was first observed through the energy loss of the Hulse-Taylor binary pulsar, and subsequent observations of that and other systems have allowed precise tests of periastron advance, Shapiro time delay, and geodetic precession. Neutron star-white dwarf binaries let us probe certain scalar-tensor theories, in which extra couplings can depend on differences in binding energy. Astronomers are tracing the orbits of stars around the supermassive black hole at the center of our galaxy, and it may soon become possible to directly image the "shadows" of black holes at the centers of our own and nearby galaxies.

Finally, cosmology is providing a rich testing ground for general relativity. Of particular interest is the question of whether the observations attributed to dark matter and dark energy are really signs of modified gravity at very large scales. While theories of modified gravity come in a wide variety, most predict differences in "structure formation"—the formation of galaxies and galactic clusters—that may be observable. Measurements of the cosmic microwave background may also lead to the detection of primordial gravitational waves, which would offer direct evidence for the quantization of gravity.

Further reading

A compilation of exact solutions to the Einstein field equations can be found in a 700-page book by Stephani et al. [117]. The post-Newtonian expansion is discussed in [64–66], and in a textbook by Poisson and Will [18]. Several textbooks now exist for numerical relativity, including Alcubierre's [118] and Baumgarte's [119].

The classic book on the global structure of spacetimes is Hawking and Ellis's *The Large Scale Structure of Space-Time* [120]. For a shorter, more informal introduction, see [121]. Reviews of some of the mathematical issues surrounding the initial value problem, singularities, stability, and global behavior can be found in Choquet-Bruhat's *General Relativity and the Einstein Equations* [122] and in Part III of the collection [123]. The topological censorship theorem was first proven in [124].

Brans-Dicke theory was introduced in [125]. A general class of "healthy" scalar-tensor theories is described in [126]; it is generalized further in [127]. A review of massive gravity is given in [128]. A more general description of alternative theories and their possible tests is found in [129].

The Wheeler-DeWitt equation was introduced in [130], as part of Bryce DeWitt's "trilogy" on covariant and canonical quantum gravity [130–132]. A good overview of quantum gravity is given in Kiefer's book [133]; see also the review [134]. Polchinski's textbook, *String Theory*, provides an excellent introduction to that topic [135]; some newer results can be found in [136]. Rovelli's textbook, *Quantum Gravity*, offers a good introduction to loop quantum gravity [137]. A nice collection of articles describing other approaches to quantum gravity can be found in [138]. For an overview of black hole thermodynamics and some of its implications for quantum gravity, see [93].

Good summaries of experimental and observational tests of general relativity can be found in [2, 4, 13]. For some interesting ideas about future tests, see [139, 140].

Appendix A
Mathematical details

This Appendix fills in a few of the more technical mathematical details that were omitted from the main text.

A.1 Manifolds

The basic definition of a manifold was introduced in Section 4.1. More formally, an n-dimensional C^k manifold M is a topological space for which:

(i) M is locally homeomorphic to $\mathbb{R}^n$; that is, each point $p \in M$ lies in an open set U that admits a homeomorphism $x\colon U \to V$, where V is an open ball in $\mathbb{R}^n$, that is, a set $\{\mathbf{x} \in \mathbb{R}^n : |\mathbf{x}|^2 < 1\}$;
(ii) In any region in which two such open sets U_1 and U_2 overlap, the transition function $x_1 \circ x_2^{-1}\colon x_2(U_1 \cap U_2) \to x_1(U_1 \cap U_2)$ is k times differentiable.

Figure A.1 illustrates this construction. Note that transition functions map $\mathbb{R}^n$ to $\mathbb{R}^n$, as required for their derivatives to make sense. (In other books, the coordinate functions, called x here, are frequently denoted φ.)

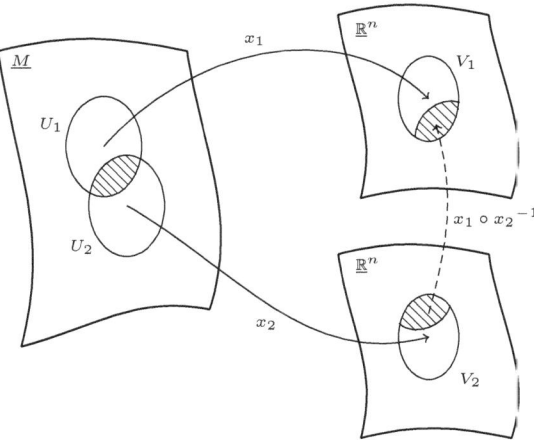

Fig. A.1 A manifold with overlapping charts

The basic definition can be generalized in various ways. For a topological manifold, the transition functions are continuous but not necessarily differentiable; for a smooth

(C^∞) manifold, they are infinitely differentiable; for an analytic manifold, they are real analytic. Further conditions are sometimes imposed on the point-set topology of the underlying space M—for instance, M may be required to be Hausdorff and paracompact or second countable—but conventions are not entirely standard.

We can also generalize to allow for the presence of boundaries. Just as a small region of manifold looks like an open subset of $\mathbb{R}^n$, a small region of manifold with boundary looks like either an open subset of $\mathbb{R}^n$ or a subset of $\mathbb{R}^n$ with an $(n-1)$-dimensional boundary. More precisely, a manifold with boundary is a topological space consisting of interior points and boundary points. Interior points lie in neighborhoods homeomorphic to open balls in $\mathbb{R}^n$, while boundary points lie in neighborhoods homeomorphic to "half balls" $\{\mathbf{x} \in \mathbb{R}^n : |\mathbf{x}|^2 < 1, x^1 \geq 0\}$, with each boundary point of M mapped to a boundary point of the corresponding half ball (a point with $x^1 = 0$). The boundary of a manifold M is usually denoted ∂M.

A triple $(U \subset M, V \subset \mathbb{R}^n, x\colon U \to V)$ is called a chart. Borrowing from map-making, a collection of charts for which every point in M is in at least one chart is called an atlas. Two atlases are equivalent if their union is also an atlas. The maximal atlas, the atlas containing all charts, is unique.

We can now give a more rigorous definition of the tangent vectors introduced in Section 4.2. A parametrized curve γ in M is a map from $\mathbb{R}$ to M. As in Chapter 2, we can choose a parameter σ as a coordinate for $\mathbb{R}$, so $\gamma(\sigma) \in M$. Let h be an arbitrary function from M to $\mathbb{R}$. Then the composition $h \circ \gamma$ is a map from $\mathbb{R}$ to $\mathbb{R}$. The tangent vector to γ at a point p is the derivative along γ acting on such functions,

$$v_p(h) = \frac{d}{d\sigma} h \circ \gamma \bigg|_{\gamma(\sigma)=p}. \tag{A.1}$$

Within a coordinate patch, we can write

$$h \circ \gamma = (h \circ x^{-1}) \circ (x \circ \gamma), \tag{A.2}$$

where $x \circ \gamma$ is now a curve in $\mathbb{R}^n$ and $h \circ x^{-1}$ is a map from $\mathbb{R}^n$ to $\mathbb{R}$, a "coordinate representative" of h. Then by the chain rule,

$$v_p(h) = \frac{d(x \circ \gamma)^\mu}{d\sigma} \frac{\partial}{\partial x^\mu} h \circ x^{-1} \bigg|_{\gamma(\sigma)=p}. \tag{A.3}$$

Equation (4.1) is shorthand for this more formal expression.

A.2 Maps between manifolds

Consider an m-dimensional manifold M and an n-dimensional manifold N, and a function $f\colon M \to N$ between them. If (U, V, x) is a chart for M and $(\tilde{U}, \tilde{V}, \tilde{x})$ is a chart for N, then the composition $\tilde{x} \circ f \circ x^{-1}$ is a map from $V \subset \mathbb{R}^m$ to $\tilde{V} \subset \mathbb{R}^n$, called a local representative, or coordinate representative, of f. The function f between manifolds is said to be k times differentiable, or C^k, if all of its local representatives are C^k. If f is one-to-one and onto and both f and f^{-1} are C^k, then f is called a C^k diffeomorphism, and M and N are said to be diffeomorphic. Note that to be diffeomorphic, two manifolds must have the same dimension.

A differentiable function between manifolds can be used to move certain tensors from one manifold to the other. Given a function $f\colon M \to N$, a tangent vector v on M can be "pushed forward" to a tangent vector f_*v on N. To define this push-forward, recall that as a tangent vector on N, f_*v should be determined by its action on functions $h\colon N \to \mathbb{R}$. We take this action to be

$$(f_*v)(h) = v(h \circ f), \tag{A.4}$$

where the right-hand side makes sense because $h \circ f$ is a function from M to $\mathbb{R}$.

A cotangent vector, on the other hand, can be "pulled back" from N to M. Recall that a cotangent vector ω on N is defined by its pairing with tangent vectors. Again given a function $f\colon M \to N$, we define the pull-back $f^*\omega$ to be the cotangent vector on M determined by the pairing

$$(f^*\omega, v) = (\omega, f_*v). \tag{A.5}$$

The push-forward operation extends in an obvious way to other type $(k, 0)$ tensors, and the pull-back extends to other type $(0, \ell)$ tensors. Tensors of a mixed type, on the other hand, can normally be neither pushed forward nor pulled back. An exception occurs, though, if f is a diffeomorphism. In that case, we can use f to push forward tangent vectors and pull back cotangent vectors, and f^{-1} to push forward cotangent vectors and pull back tangent vectors, with an obvious generalization to tensors of any type. Thus a diffeomorphism does more than merely map one manifold to another; it also carries along all of the manifold's geometric and physical structure.

Diffeomorphisms play a central role in general relativity. They are the fundamental symmetries of the theory, the "active" coordinate transformations. Recall that in ordinary flat space, a "passive" rotation is a rotation of the coordinate axes with all physical objects remaining fixed, while an "active" rotation is a rotation of the physical world with the coordinate axes remaining fixed. Although conceptually very different, in practice these are equivalent in any rotationally invariant theory. In the same way, in a diffeomorphism-invariant theory like general relativity, "active" diffeomorphisms from M to itself are equivalent to "passive" coordinate transformations, as long as all geometric and physical objects are carried along—pushed forward—in the process. The collection of diffeomorphisms from M to itself, $\mathit{Diff}(M)$, forms a group under the composition of functions. This group is the "gauge group" of general relativity; two diffeomorphic spacetimes are considered to be two different descriptions of the same physics. This is the formal expression of the "principle of general covariance" of Section 4.1.

A.3 Topologies

The topological properties of a manifold are those properties that are preserved under continuous deformations. Two manifolds M and N are said to have the same topology if there exists a homeomorphism $f\colon M \to N$ between them. The topology of manifolds is an enormous topic, widely studied in mathematics, and for the most part it lies outside the scope of an introductory book like this one. Still, it is useful to have in mind a few examples of interesting topologies, some of which may be relevant to our Universe (see Box 11.1).

120   *Mathematical details*

- **The torus:**
  - In one dimension, start with a line segment, and identify ("glue together") the end points. The resulting circle is topologically a one-dimensional torus T^1. (The superscript denotes the dimension of the manifold.)
  - In two dimensions, start with a square, or more generally a parallelogram, and identify opposite edges (without twisting). The resulting "donut" is a two-dimensional torus T^2. This is sometimes called the "video game model" of the torus, for its resemblance to a video game in which objects disappear out of one side of the screen and instantly reappear on the opposite side.
  - In three dimensions, start with a cube, and identify opposite faces to form a three-torus T^3. The same construction in d dimensions yields a d-torus T^d.

  These manifolds can be hard to visualize—even the three-torus doesn't "fit" into ordinary three-dimensional Euclidean space—but we can still describe their properties. In three dimensions, for example, a line perpendicular to one face of the cube will travel through the cube, reach the opposite edge, and "jump" back to its starting point to form a closed curve. In general, a d-torus has d such "circumferences," each an independent closed curve that wraps around the manifold and cannot be continuously shrunk to a point.

- **The sphere:**
  - In one dimension, start with two line segments and identify the boundary of one (that is, the two end points) with the boundary of the other. The result is again a circle, which is both a one-dimensional torus and a one-dimensional sphere S^1.
  - In two dimensions, start with two disks (or "hemispheres") and identify the boundary of one with the boundary of the other. The resulting space is a two-sphere S^2.
  - In three dimensions, start with two (solid) balls and identify their boundaries to obtain a three-sphere S^3. The same construction in d dimensions gives the d-sphere S^d.

  Again, these manifolds are hard to visualize, but have properties we can describe. For instance, in two dimensions a great circle encircles a sphere and returns to its starting point. In three dimensions the same is true: a straight line starting at the center of one ball will reach the edge, "jump" to the identified point on the second ball, travel through the center and out the opposite side, "jump" back to the first ball, and return to its starting point.

- **Other identifications:**

  For the sphere and the torus, we started with simple pieces of ordinary $\mathbb{R}^d$ and built more complicated topologies by "gluing together" their boundaries. The same method can be used for many other topologies. In two dimensions, a genus g surface (a "g-holed donut") can be built by identifying pairs of edges of a $4g$-sided polygon. Infinitely many topologically distinct manifolds with constant curvature metrics can be built by taking pieces of the three-sphere or three-dimensional hyperbolic space, with the metrics (11.4), and identifying faces by means of discrete groups of isometries. In three dimensions, in fact, it has recently been proven that any topological manifold can be built by appropriately gluing together pieces of spaces with a few simple geometries.

The notion of topology has various generalizations. If two manifolds M and N are homeomorphic but not diffeomorphic, for instance, they are said to have the same topology but different differential structures. In three or fewer dimensions, a topological manifold has a unique differential structure, but in four dimensions the number of distinct differential structures can be infinite. In fact, $\mathbb{R}^d$ has a unique differential structure for $d \neq 4$, but, intriguingly, $\mathbb{R}^4$ has infinitely many "exotic" differential structures, a feature that may yet prove interesting in physics.

A.4 Cohomology

A basic problem in topology is to find topological invariants, functions that depend only on the topology of a manifold. Such invariants can be used to characterize topology and to distinguish manifolds with different topologies. The Cartan calculus of Section 5.3 makes it possible to construct a widely used set of invariants, the de Rham cohomology groups.

A p-form α on a manifold M is said to be closed if $d\alpha = 0$, and exact if $\alpha = d\beta$ for some $(p-1)$-form β. Every exact form is closed, since $d^2 = 0$. Every closed form is *locally* exact: if $d\alpha = 0$, then given any point in the manifold, there is a $(p-1)$-form β defined in a neighborhood of that point for which $\alpha = d\beta$. But this local result may not extend over the whole manifold; whether or not it does depends on the topology of M.

As the simplest example, consider the one-form $\alpha = d\theta$ on a circle. This form is certainly closed, and it looks as if it is exact, the exterior derivative of a zero-form θ. But θ is not single-valued, and is not globally defined on the circle. We need at least two coordinate patches to cover a circle, and while $d\theta$ is a tensor that can be defined independent of the coordinate choice, θ is not. We shall see in the next section that if ω is an exact one-form, its integral over a closed one-dimensional manifold must vanish, while here, $\int_{S^1} d\theta = 2\pi$. This example extends easily to the n-torus T^n, which can be described by a set of periodic coordinates $\{x^1, x^2, \ldots, x^n\}$ with $x^i \sim x^i + a^i$ for some constants a^i; now each of the one-forms dx^i is closed but not exact.

Let $Z^p(M)$ be the space of closed p-forms on M, and $B^p(M)$ be the space of exact p-forms, both viewed as vector spaces. Define

$$H^p(M) = Z^p(M)/B^p(M), \qquad (A.6)$$

where the quotient means that closed forms are considered equivalent if they differ by an exact form,

$$\alpha_1 \sim \alpha_2 \text{ if } \alpha_2 - \alpha_1 = d\beta \text{ for some } \beta. \qquad (A.7)$$

$H^p(M)$ is the p-th de Rham cohomology group of M, and its dimension is the p-th Betti number. It is easy to see that these objects are invariant under diffeomorphisms, which take closed forms to closed forms and exact forms to exact forms. The de Rham cohomology groups do not completely characterize the topology—two topologically inequivalent manifolds can have the same de Rham cohomology—but they provide a partial characterization: if $H^p(M) \neq H^p(N)$, then M and N are topologically distinct.

A.5 Integrals of d-forms and Stokes' theorem

We saw in Section 5.1 that if we have a metric, we can integrate a quantity of the form $f\sqrt{|g|}$, where f is a scalar. In the absence of a metric, integrals can still be defined; on a d-dimensional manifold, the objects that can be integrated are d-forms.

Let $\omega_{\mu_1\mu_2...\mu_d}$ be the components of a d-form in a coordinate basis. By definition, ω is totally antisymmetric, so in d dimensions it has only one independent component, which we can call $\hat{\omega}$:

$$\omega_{\mu_1\mu_2...\mu_d} = \hat{\omega}\,\tilde{\epsilon}_{\mu_1\mu_2...\mu_d}\,, \tag{A.8}$$

where $\tilde{\epsilon}_{\mu_1\mu_2...\mu_d}$ is the Levi-Civita symbol (4.19). Transforming to new coordinates $\bar{x}^\mu$,

$$\bar{\omega}_{\nu_1\nu_2...\nu_d} = \hat{\omega}\,\tilde{\epsilon}_{\mu_1\mu_2...\mu_d} \left(\frac{\partial x^{\mu_1}}{\partial \bar{x}^{\nu_1}}\right) \left(\frac{\partial x^{\mu_2}}{\partial \bar{x}^{\nu_2}}\right) \cdots \left(\frac{\partial x^{\mu_d}}{\partial \bar{x}^{\nu_d}}\right). \tag{A.9}$$

But it is a basic fact of linear algebra that if $M^\mu{}_\nu$ is any $d \times d$ matrix,

$$\tilde{\epsilon}_{\mu_1\mu_2...\mu_d}\,M^{\mu_1}{}_{\nu_1} M^{\mu_2}{}_{\nu_2} \cdots M^{\mu_d}{}_{\nu_d} = \det|M|\,\tilde{\epsilon}_{\nu_1\nu_2...\nu_d}\,. \tag{A.10}$$

Comparing (A.9), we see that

$$\bar{\hat{\omega}} = \hat{\omega}\,J^{-1} \quad \text{with } J = \det\left|\frac{\partial \bar{x}}{\partial x}\right|, \tag{A.11}$$

where J is the Jacobian (5.2). Hence $\hat{\omega}$, like $\sqrt{|g|}$, is a scalar density of weight -1, and the combination $\hat{\omega}\,d^d x$ is an invariant, which can be integrated.

More generally, an object that transforms as

$$\bar{X} = X\,J^w \tag{A.12}$$

is called a scalar density of weight w. A scalar density of weight w times a type (k, ℓ) tensor is a type (k, ℓ) tensor density of weight w. The Levi-Civita symbol $\tilde{\epsilon}_{\mu_1\mu_2...\mu_d}$ can be viewed as a type $(0, d)$ tensor density of weight 1, since such a tensor density transforms as

$$\bar{\tilde{\epsilon}}_{\mu_1\mu_2...\mu_d} = \tilde{\epsilon}_{\mu_1\mu_2...\mu_d} \left(\frac{\partial x^{\mu_1}}{\partial \bar{x}^{\nu_1}}\right) \left(\frac{\partial x^{\mu_2}}{\partial \bar{x}^{\nu_2}}\right) \cdots \left(\frac{\partial x^{\mu_d}}{\partial \bar{x}^{\nu_d}}\right) J = \tilde{\epsilon}_{\mu_1\mu_2...\mu_d}\,. \tag{A.13}$$

The combination

$$\epsilon_{\mu_1\mu_2...\mu_d} = \sqrt{|g|}\,\tilde{\epsilon}_{\mu_1\mu_2...\mu_d} \tag{A.14}$$

is a genuine type $(0, d)$ tensor, the Levi-Civita tensor.

The Levi-Civita tensor allows us to define a new operation on differential forms, the Hodge dual or Hodge star. Let ω be a p-form on a d-dimensional manifold. Its Hodge dual is the $(d-p)$-form $*\omega$ defined by

$$*\omega_{a_1 a_2 \ldots a_{d-p}} = \frac{1}{p!} \epsilon_{a_1 a_2 \ldots a_{d-p}}{}^{b_1 b_2 \ldots b_p} \omega_{b_1 b_2 \ldots b_p} \,, \tag{A.15}$$

where the prefactor is chosen so that $**\omega = \pm \omega$. (For a p-form in a four-dimensional Lorentzian spacetime, the sign is $(-1)^{p+1}$.) The Hodge dual $*\omega$ has the same number of independent components as ω, and contains the same information. A familiar example from three dimensions is the cross-product: the antisymmetric product of two vectors is "really" a two-form, and the familiar expression is its Hodge dual.

We can now define a new differential operator, the codifferential $\delta = *d*$, which takes p-forms to $(p-1)$-forms. Like d, δ satisfies the identity $\delta^2 = 0$. It is not hard to show that the combination $\Delta = d\delta + \delta d$ is the Laplacian on forms. The generalization of the Helmholtz decomposition—the theorem that a three-vector $\mathbf{v}$ can always be written as $\mathbf{v} = \nabla \phi + \nabla \times \mathbf{w}$—is the Hodge decomposition,

$$\omega = d\alpha + \delta\beta + \gamma \quad \text{with } \Delta\gamma = 0 \,. \tag{A.16}$$

A form for which $\Delta\gamma = 0$ is called a harmonic form.

A number of results from vector calculus—Green's theorem, Stokes' theorem, the divergence theorem—relate integrals over a manifold to integrals over its boundary. In the language of differential forms, these all take the same simple form. Let M be a d-dimensional manifold with boundary ∂M, and let ω be a $(d-1)$-form on M. Then the d-form $d\omega$ can be integrated over M, and Stokes' theorem is simply that

$$\int_M d\omega = \int_{\partial M} \omega \,. \tag{A.17}$$

In particular, if α is a one-form, its Hodge dual $*\alpha$ is a $(d-1)$-form, and

$$\int_M d*\alpha = \int_{\partial M} *\alpha \,. \tag{A.18}$$

It is not hard to check that the left-hand side of (A.18) is

$$\int_M \partial_\mu (\sqrt{|g|} \alpha^\mu) \, d^d x \,,$$

yielding the divergence theorem and allowing integration by parts as in Section 5.5.

A.6 Curvature and holonomies

Section 5.7 introduced a definition of curvature in terms of parallel transport around a curve. Here we fill in some of the details.

Start with a parametrized curve γ, described in a coordinate representation by $x^\mu(\sigma)$, with the parameter σ normalized to $\sigma \in [0,1]$. From Section 5.6, the change of a vector v^a under parallel transport along γ is

$$\frac{dx^\mu}{d\sigma}\nabla_\mu v^a = 0 = \frac{dv^a}{d\sigma} + A^a{}_b v^b \quad \text{with } A^a{}_b = \Gamma_\mu{}^a{}_b \frac{dx^\mu}{d\sigma}. \tag{A.19}$$

We can convert this to an integral equation,

$$v^a(\sigma) = v^a(0) - \int_0^\sigma d\sigma_1\, A^a{}_b(\sigma_1) v^b(\sigma_1), \tag{A.20}$$

and by iteration,

$$v^a(\sigma) = v^a(0) - \int_0^\sigma d\sigma_1\, A^a{}_b(\sigma_1)\Big(v^b(0)$$
$$- \int_0^{\sigma_1} d\sigma_2\, A^b{}_c(\sigma_2)\Big(v^c(0) - \int_0^{\sigma_2} d\sigma_3\, A^c{}_d(\sigma_3)\Big(v^d(0) - \ldots\Big. \tag{A.21}$$

To put this into a more elegant form, note that for any matrix X,

$$\int_0^\sigma d\sigma_1 \int_0^{\sigma_1} d\sigma_2\, X(\sigma_1)X(\sigma_2) = \iint_{\sigma_2<\sigma_1<\sigma} d\sigma_1 d\sigma_2\, X(\sigma_1)X(\sigma_2) \tag{A.22}$$

$$= \frac{1}{2!}\iint_{\sigma_2<\sigma_1<\sigma} d\sigma_1 d\sigma_2\, X(\sigma_1)X(\sigma_2) + \frac{1}{2!}\iint_{\sigma_1<\sigma_2<\sigma} d\sigma_1 d\sigma_2\, X(\sigma_2)X(\sigma_1),$$

where the two terms on the last line are equal. Now define a new operation on matrices, the path ordering operation P, by

$$P(X(\sigma_1)Y(\sigma_2)) = \begin{cases} X(\sigma_1)Y(\sigma_2) & \text{if } \sigma_1 > \sigma_2 \\ Y(\sigma_2)X(\sigma_1) & \text{if } \sigma_2 > \sigma_1 \end{cases}. \tag{A.23}$$

(The mnemonic is "later on the left.") Then (A.22) becomes

$$\int_0^\sigma d\sigma_1 \int_0^{\sigma_1} d\sigma_2\, X(\sigma_1)X(\sigma_2) = \frac{1}{2!}\int_0^\sigma d\sigma_1 \int_0^\sigma d\sigma_2\, P(X(\sigma_1)X(\sigma_2)). \tag{A.24}$$

The same argument can be applied to products of more than two matrices, where a product of n matrices has $n!$ possible orderings. If we now treat the components $A^a{}_b$ in (A.21) as matrices, we can rewrite the equation as

$$v^a(\sigma) = P\Big(\delta^a_b - \int_0^\sigma d\sigma_1\, A^a{}_b(\sigma_1) + \frac{1}{2!}\int_0^\sigma d\sigma_1 \int_0^\sigma d\sigma_2\, (A(\sigma_1)A(\sigma_2))^a{}_b$$
$$- \frac{1}{3!}\int_0^\sigma d\sigma_1 \int_0^\sigma d\sigma_2 \int_0^\sigma d\sigma_3\, (A(\sigma_1)A(\sigma_2)A(\sigma_3))^a{}_b + \ldots\Big) v^b(0). \tag{A.25}$$

This expression closely resembles the Taylor expansion of an exponential, and is called a path-order exponential, abbreviated as

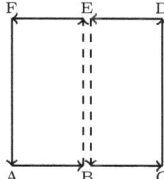

Fig. A.2 A large contractable curve may be divided into smaller ones

$$v^a(\sigma) = P\exp\left\{-\int_0^\sigma d\sigma_1\, A(\sigma_1)^a{}_b\right\} v^b(0) = P\exp\left\{-\int_\gamma \Gamma_\mu{}^a{}_b\, dx^\mu\right\} v^b(0). \quad (A.26)$$

This form may be familiar from quantum field theory: it is essentially the same as the expression for the S-matrix as a time ordered product. The matrix

$$U^a{}_b(\sigma,\sigma') = P\exp\left\{-\int_\sigma^{\sigma'} d\sigma_1 A(\sigma_1)^a{}_b\right\} \quad (A.27)$$

is called the parallel transport matrix. For a closed curve, one for which $x^\mu(0) = x^\mu(1)$, $U^a{}_b(0,1) = H^a{}_b$ is the holonomy of the curve. Such holonomies are the starting point for the observables in loop quantum gravity [137].

Let us now consider a very small closed curve γ, of length $\ell \ll 1$. Each new integral in (A.25) brings in a factor of the length of the curve, so as long as the connection is bounded, the first terms should dominate. There is one exception, though: by Stokes' theorem, the first nontrivial term can be reexpressed as a double integral,

$$\int_\gamma \Gamma_\mu{}^a{}_b dx^\mu = \int_\Sigma (\partial_\mu \Gamma_\nu{}^a{}_b - \partial_\nu \Gamma_\mu{}^a{}_b) dx^\mu dx^\nu, \quad (A.28)$$

where Σ is any surface bounded by the curve γ. The terms with one and two integrals in (A.27) are thus of the same order, and, in fact, combine neatly to give the curvature tensor,

$$H^a{}_b = \delta^a_b + \int_\Sigma R_{\mu\nu}{}^a{}_b\, d\sigma^{\mu\nu} + \mathcal{O}(\ell^3). \quad (A.29)$$

Equation (A.29) holds for very small curves. But the curvature is often sufficient to determine the holonomies of large curves as well. Imagine transporting a vector around the path BCDEFAB in Fig. A.2. We can split the path into two pieces, BCDEB followed by BEFAB. The contributions of BE and EB cancel, and it is straightforward to see that the holonomy of the larger curve is simply the product of the two smaller holonomies. Each smaller curve may in turn be subdivided, and the total holonomy ultimately becomes a product of holonomies of infinitesimal curves, for which the higher order terms in (A.29) can be neglected.

There is one important exception, though. We have implicitly assumed that the curve BCDEFAB encloses a surface that can be subdivided. For a topologically nontrivial manifold, this may not be true. A circumference of a torus, for example, encloses no surface, and can have holonomies unrelated to the curvature. In fact, a flat torus, one with a curvature tensor that vanishes everywhere, can still have nonzero holonomies.

The relevant topological property here is the existence of noncontractible curves, curves that cannot be shrunk continuously to a point. The equivalence classes of such curves form a group, the "fundamental group" $\pi_1(M)$, which is closely related to the first cohomology group $H^1(M)$ discussed in Section A.4. The full geometry of a topologically nontrivial manifold M is determined by the curvature tensor plus the holonomies around a set of generators of $\pi_1(M)$.

A.7 Symmetries and Killing vectors

Sections 10.1–10.2 introduced the symmetries of the Schwarzschild spacetime in a rather coordinate-dependent way. For a more geometric approach, we can define a symmetry—technically an isometry—by demanding that a shift $x^\mu \to x^\mu + \xi^\mu$ by some vector field ξ leave the metric (as a tensor!) unchanged. By (4.32), ξ must be a Killing vector,

$$\delta_\xi g_{\mu\nu} = g_{\mu\rho}\partial_\nu \xi^\rho + g_{\nu\rho}\partial_\mu \xi^\rho + \xi^\rho \partial_\rho g_{\mu\nu} = 0 \Leftrightarrow \nabla_\mu \xi_\nu + \nabla_\nu \xi_\mu = 0. \qquad (A.30)$$

Given such a Killing vector, though, we can change to "adapted coordinates" in which the coordinate-dependent definitions make sense. Suppose, for example, that ξ is timelike, so the metric is invariant under translations in a timelike direction. Choose new coordinates for which $\bar{\xi}^0 = 1$ and $\bar{\xi}^i = 0$. This is always possible; we simply have to solve the equations

$$\bar{\xi}^0 = \xi^\rho \frac{\partial \bar{x}^0}{\partial x^\rho} = 1, \quad \bar{\xi}^i = \xi^\rho \frac{\partial \bar{x}^i}{\partial x^\rho} = 0 \qquad (A.31)$$

for $\bar{x}^\mu$. In these coordinates, the Killing equation (A.30) reduces to $\partial_{\bar{0}} \bar{g}_{\mu\nu} = 0$, so it now makes sense to say that the components of the metric are independent of time.

Intuitively, the vector field ξ defines a flow, whose lines of flow, or streamlines, are the integral curves

$$\frac{dx^\mu}{ds} = \xi^\mu(x(s)). \qquad (A.32)$$

Then

$$\xi^\rho \partial_\rho = \frac{d}{ds}, \qquad (A.33)$$

and (A.31) is simply the statement that the $\bar{x}^i$ are constant on each streamline and $\bar{x}^0$ is an affine parameter. The added condition for a static metric can be shown to be equivalent to the requirement that these streamlines are hypersurface orthogonal; the coordinate-free statement is that $\xi_{[a}\nabla_b \xi_{c]} = 0$.

Similarly, the geometric condition for spherical symmetry is the existence of three Killing vectors that generate the group of rotations SO(3). This condition may be familiar from quantum mechanics, where angular momentum is the generator of rotations; the requirement is essentially that the Killing vectors have the same commutators as the components of angular momentum. A corresponding choice of adapted coordinates will then reproduce the metric (10.2).

More generally, even if ξ is not a Killing vector field, we can define integral curves (A.32) and study the change of a tensor along these curves. This change is known as the Lie derivative, denoted by $\mathcal{L}_\xi$, and can be used to describe the response of a tensor to a diffeomorphism. Equation (4.31) is an example of such a construction; it may be reexpressed in the form

$$\mathcal{L}_\xi g_{\mu\nu} = \nabla_\mu \xi_\nu + \nabla_\nu \xi_\mu . \tag{A.34}$$

Lie derivatives of other tensors are defined similarly, where, as in Section 4.6, one must keep track of changes of both the components and the basis. Like the exterior derivative, the Lie derivative does not require a connection, but it is more "nonlocal": the Lie derivative with respect to a vector field ξ depends not only on ξ, but also on derivatives of ξ.

A.8 The York decomposition

Section 12.5 briefly discussed the York decomposition and the description of "free data" for solutions of the Einstein field equations. This section fills in a few details.

Let us restrict ourselves to spacetimes with constant mean curvature (CMC) time slicings, as described in Section 12.5. That is, we assume our spacetime admits a time coordinate for which each hypersurface of constant time has constant mean curvature $K = q^{ij} K_{ij}$, where K_{ij} is the extrinsic curvature (12.6). If it exists, such a slicing is normally unique, so we can use K itself as a time coordinate, $K = -t$. Not every spacetime admits a CMC slicing, but at this writing the known exceptions seem rather contrived. Some of the technical results of this section also require the slices Σ to be compact, although generalizations are available.

On each time slice, the spatial metric q_{ij} is conformal to a "Yamabe metric," a metric whose scalar curvature is constant. That is, there is always a function ϕ such that

$$q_{ij} = \phi^4 \bar{q}_{ij}, \text{ with } {}^{(3)}R[\bar{q}] = 0 \text{ or } \pm 1, \tag{A.35}$$

where the constant is determined by the topology of Σ. We can also form a "transverse traceless decomposition" of the conjugate momentum, writing

$$\pi^{ij} = \phi^{-4}\left(p^{ij} + (PY)^{ij}\right) + \frac{1}{3}q^{ij}\pi, \text{ with } \bar{\nabla}_i p^{ij} = 0, \; \bar{q}_{ij} p^{ij} = 0,$$

$$\text{and } (PY)^{ij} = \bar{q}^{ik}\bar{\nabla}_k Y^j + \bar{q}^{jk}\bar{\nabla}_k Y^i - \frac{2}{3}\bar{q}^{ij}\bar{\nabla}_k Y^k, \tag{A.36}$$

where $\bar{\nabla}$ is the covariant derivative compatible with the metric $\bar{q}_{ij}$. The momentum p^{ij} is the "transverse traceless" piece, $(PY)^{ij}$ is traceless but not transverse, and from (12.10), the trace is $\pi = \frac{1}{\kappa^2}\sqrt{q}\,K = -\frac{1}{\kappa^2}\phi^6\sqrt{\bar{q}}\,t$.

It is now straightforward to check that the momentum constraint $\mathcal{H}^i = 0$ reduces to the requirement that $\bar{\nabla}_i (PY)^{ij} = 0$, which in turn forces $(PY)^{ij}$ to vanish. The Hamiltonian constraint (12.13) then becomes

$$\mathcal{H} = \frac{4}{\kappa^2}\phi\sqrt{\bar{q}}\left[\bar{\Delta}\phi - \frac{1}{8}\phi\bar{R} + \frac{\kappa^4}{2\bar{q}}\phi^{-7}\bar{q}_{ij}\bar{q}_{kl}p^{il}p^{jk} - \frac{1}{12}\phi^5 t^2\right] = 0, \tag{A.37}$$

where $\bar{\Delta} = \bar{q}^{ij}\bar{\nabla}_i\bar{\nabla}_j$. Equation (A.37), known as the Lichnerowicz-York equation, determines ϕ in terms of the free data $(\bar{q}_{ij}, p^{ij})$. It is an elliptic differential equation with well-understood properties, and normally has a unique solution. Given such a solution, we can use (A.35) and (A.36) to reconstruct initial data that obey the constraints, and then (in principle) evolve that data to find a full solution of the Einstein field equations. In practice, the solution of the Lichnerowicz-York equation can very rarely be written in closed form, but the formalism provides an important starting point for numerical relativity.

We can also insert (A.35) and (A.36) back into the action to find an effective Hamiltonian for our free data. Since the constraints have been solved, the action (12.14) reduces to

$$I_{\text{grav}} = \int dt \int_\Sigma d^3x \, \pi^{ij} \partial_0 q_{ij} = \int dt \int_\Sigma d^3x \left(p^{ij} \partial_0 \bar{q}_{ij} + \frac{1}{3} \pi q^{ij} \partial_0 q_{ij} \right)$$
$$= \int dt \int_\Sigma d^3x \left(p^{ij} \partial_0 \bar{q}_{ij} + \frac{2}{3\kappa^2} \phi^6 \sqrt{\bar{q}} \right), \quad (A.38)$$

where the final step comes from integration by parts and the equality $\pi = -\frac{1}{\kappa^2}\sqrt{\bar{q}}\,t$. This is an example of a "reduced phase space" action, with unconstrained variables and an ordinary Hamiltonian. The Hamiltonian, though, depends on ϕ, which is determined by solving the Lichnerowicz-York equation, and is thus a highly nonlocal function of the phase space variables.

Further reading

Much of the material in this appendix is discussed in references from Chapters 5 and 6, in particular [32, 33, 35, 41]. Ref. [34] has a very readable informal introduction to manifold topologies; for a more technical introduction to the geometry and topology of three-manifolds, see Thurston's *Three-Dimensional Geometry and Topology* [141]. The proof that any topological three-manifold can be obtained by gluing simple geometrical pieces is due to Perelman; for an overview, see [142]. A discussion of differential structures and their possible physical relevance can be found in [143].

A further discussion of de Rham cohomology can be found in [33] and [41]. The appendices of [35] have a good discussion of integration and Stokes' theorem, along with an introduction to Frobenius's theorem, which is used in the formal definition of a static metric. Section 6.1 of [35] also has a good discussion of Killing vectors and spherical symmetry, while Chapter 8 of [117] has a much more general discussion of possible symmetries and Killing vectors. Lie derivatives are described in appendix C.2 of [35] and appendix B of [17].

A careful mathematical treatment of the Lichnerowicz-York equation can be found in chapter VII of [122]; see also the readings at the end of Chapter 12.

References

[1] Ostrogradsky, M., Mem. Ac. St. Petersbourg VI 4 (1850) 385.
[2] Will, C. M., *Theory and experiment in gravitational physics* (Cambridge University Press, Cambridge, 1993), 2nd edition.
[3] Eötvös, R. v., Pekár, D., and Fekete, E., Ann. Phys. (Leipzig) 68 (1922) 11.
[4] Will, C. M., Living Rev. Relativ. 17 (2014) 4, DOI: 10.12942/lrr-2014-4.
[5] Woodard, R. P., Scholarpedia 10 (2015) 32243, DOI: 10.4249/scholarpedia.32243.
[6] Geroch, R., *General Relativity from A to B* (University of Chicago Press, Chicago, 1978).
[7] Carlip, S. and DeWitt-Morette, C., Phys. Rev. Lett. 60 (1988) 1599, DOI: 10.1103/PhysRevLett.60.1599.
[8] d'Inverno, R., *Introducing Einstein's Relativity* (Clarendon Press, Oxford, 1992), Section 7.6.
[9] Taylor, E. F. and Wheeler, J. A., *Spacetime Physics* (W. H. Freeman and Co., New York, 1992).
[10] Schutz, B. F., *A First Course in General Relativity* (Cambridge University Press, Cambridge, 2009).
[11] Misner, C. W., Thorne, K. S., and Wheeler, J. A., *Gravitation* (Princeton University Press, Princeton, 2017).
[12] Bodenner, J. and Will, C. M., Am. J. Phys. 71 (2003) 770, DOI: 10.1119/1.1570416.
[13] Ni, W.-T., in *One Hundred Years of General Relativity: From Genesis and Empirical Foundations to Gravitational Waves, Cosmology and Quantum Gravity*, edited by Ni, W.-T. (World Scientific, Singapore, 2016), chap. 8.
[14] Will, C. M., Am. J. Phys. 78 (2010) 1240, DOI: 10.1119/1.3481700.
[15] Schwarzschild, K., Sitzungsber. Preuss. Akad. Wiss. Berlin (Math. Phys.) (1916) 189; English translation by Antoci, S. and Loinger, A., in Gen. Rel. Grav. 35 (2003) 951, DOI: 10.1023/A:1022971926521.
[16] Hackmann, E. and Lämmerzahl, C., Phys. Rev. Lett. 100 (2008) 171101, DOI: 10.1103/PhysRevLett.100.171101.
[17] Carroll, S., *Spacetime and Geometry* (Pearson Education, London, 2004).
[18] Poisson, E. and Will, C. M., *Gravity: Newtonian, Post-Newtonian, Relativistic* (Cambridge University Press, Cambridge, 2014).
[19] Roseveare, N. T., *Mercury's perihelion from Le Verrier to Einstein* (Oxford University Press, Oxford, 1982).
[20] Park, R. S., Folkner, W. M., Konopliv, A. S., Williams, J. G., Smith, D. E., and Zuber, M. T., Astron. J. 153 (2017) 121, DOI: 10.3847/1538-3881/aa5be2.
[21] Dyson, F. W., Eddington, A. S., and Davidson, C., Phil. Trans. Roy. Soc. Lon. A 220 (1920) 291, DOI: 10.1098/rsta.1920.0009.

[22] Kennefick, D., Physics Today 62 (March 2009) 37, DOI: 10.1063/1.3099578.
[23] Shapiro, S. S., Davis, J. L., Lebach, D. E., and Gregory, J. S., Phys. Rev. Lett. 92 (2004) 121101, DOI: 10.1103/PhysRevLett.92.121101.
[24] Dodelson, S., *Gravitational Lensing* (Cambridge University Press, Cambridge, 2017).
[25] Treu, T., Marshall, P. J., and Clowe, D., Am. J. Phys. 80 (2012) 753, DOI: 10.1119/1.4726204.
[26] Shapiro, I. I., Phys. Rev. Lett. 13 (1964) 789, DOI: 10.1103/PhysRevLett.13.789.
[27] Bertotti, B., Iess, L., and Tortora, P., Nature 425 (2003) 374, DOI: 10.1038/nature01997.
[28] Pound, R. V. and Rebka, G. A., Phys. Rev. Lett. 4 (1960) 337, DOI: 10.1103/PhysRevLett.4.337.
[29] Chou, C. W., Hume, D. B., Rosenband, T., and Wineland, D. J., Science 329 (2010) 1630, DOI: 10.1126/science.1192720.
[30] Ashby, N., Living Rev. Relativ. 6 (2003) 1, DOI: 10.12942/lrr-2003-1.
[31] Wheeler, J. A., *A Journey into Gravity and Spacetime* (Scientific American Library, New York, 1999).
[32] Isham, C. J., *Modern Differential Geometry for Physicists* (World Scientific, Singapore, 1999).
[33] Choquet-Bruhat, Y. and DeWitt-Morette, C., *Analysis, Manifolds and Physics* (Elsevier, Amsterdam, 1982).
[34] Weeks, J. R., *The Shape of Space* (Marcel Dekker, New York, 2002).
[35] Wald, R. M., *General Relativity* (University of Chicago Press, Chicago, 1984).
[36] Cartan, H., *Differential Forms* (Dover Publications, Minneola, 1970).
[37] Kretschmann, E., Ann. Phys. 358 (1918) 575, DOI: 10.1002/andp.19183581602.
[38] Anderson, J. L., *Principles of Relativity Physics* (Academic Press, San Diego, 1967).
[39] Ehlers, J., Pirani, F. A. E., and Schild, A., in *General Relativity: papers in honour of J. L. Synge*, edited by O'Raifeartaigh, L. (Clarendon Press, Oxford, 1972); reprinted in Gen. Rel. Grav. 44 (2012) 1587, DOI: 10.1007/s10714-012-1353-4.
[40] Cartan, E., Ann. Éc. Norm. 40 (1923) 325.
[41] Eguchi, T., Gilkey, P. B., and Hanson, A. J., Phys. Rept. 66 (1980) 213, DOI: 10.1016/0370-1573(80)90130-1.
[42] Poisson, E., *A Relativist's Toolkit* (Cambridge University Press, Cambridge, 2004).
[43] Noether, E., Nachr. Gesellsch. Wiss. Göttingen (Math. Phys.) 1918 (1918) 235.
[44] Einstein, A., Sitzungsber. Preuss. Akad. Wiss. Berlin (Math. Phys.) 1915 (1915) 844.
[45] Einstein, A., Ann. Phys. 354 (1916) 769, DOI: 10.1002/andp.19163540702; English translation by Perrett, W. and Jeffrey, G. B. in *The Principle of Relativity* (Dover, Toronto, 1952), chap. 7.
[46] Hilbert, D., Nach. Ges. Wiss. Göttingen 1915 (1915) 395.
[47] Vermeil, H., Nach. Ges. Wiss. Göttingen 1917 (1917) 334.
[48] Lovelock, D., J. Math. Phys. 12 (1971) 498, DOI: 10.1063/1.1665613.

[49] Boulware, D. G. and Deser, S., Ann. Phys. 89 (1975) 193, DOI: 10.1016/0003-4916(75)90302-4.
[50] Jacobson, T., Phys. Rev. Lett. 75 (1995) 1260, DOI: 10.1103/PhysRevLett.75.1260.
[51] Corry, L., Renn, J., and Stachel, J., Science 278 (1997) 1270, DOI: 10.1126/science.278.5341.1270.
[52] Weinberg, S., *The Quantum Theory of Fields*, Vol. I (Cambridge University Press, Cambridge, 1995).
[53] Palatini, A., Rend. Circ. Mat. Palermo 43 (1919) 203; English translation by Hojman, R. and Mukku, C. in *Cosmology and Gravitation*, edited by Bergmann, P. G. and DeSabbata, V. (Plenum Press, New York, 1980).
[54] Dadhich, N. and Pons, J. M., Gen. Rel. Grav. 44 (2012) 2337, DOI: 10.1007/s10714-012-1393-9.
[55] Corichi, A., Rubalcava-García, I., and Vukašinac, T., Int. J. Mod. Phys. D25 (2016) 1630011, DOI: 10.1142/S0218271816300111.
[56] Hehl, F. W., von der Heyde, P., Kerlick, G. D., and Nester, J. M., Rev. Mod. Phys. 48 (1976) 393, DOI: 10.1103/RevModPhys.48.393.
[57] Starobinsky, A. A., Phys. Lett. 91B (1980) 99, DOI: 10.1016/0370-2693(80)90670-X.
[58] Norton, J. D., in *The Genesis of General Relativity*, edited by Janssen, M., Norton, J. D., Renn, J., Sauer, T., and Stachel, J., Boston Stud. Phil. Sci. vol 250 (Springer, Dordrecht, 2007), p. 1337.
[59] Lord, E. A., J. Math. Phys. 17 (1976) 37, DOI: 10.1063/1.522800.
[60] de Felice, F. and Clarke, C. J. S., *Relativity on curved manifolds* (Cambridge University Press, Cambridge, 1990).
[61] Brown, J. D. and York, J. W., Phys. Rev. D47 (1993) 1407, DOI: 10.1103/PhysRevD.47.1407.
[62] Szabados, L. B., Living Rev. Relativ. 12 (2009) 4, DOI: 10.12942/lrr-2009-4.
[63] Einstein, A., Infeld, L., and Hoffmann, B., Annals Math. 39 (1938) 65, DOI: 10.2307/1968714.
[64] Damour, T., in *Three Hundred Years of Gravitation*, edited by Hawking, S. and Israel, W. (Cambridge University Press, Cambridge, 1987), chap. 6.
[65] Blanchet, L., Living Rev. Relativ. 5 (2002) 3, DOI: 10.12942/lrr-2002-3.
[66] Maggiore, M., *Gravitational Waves: Volume 1: Theory and Experiments* (Oxford University Press, Oxford, 2008).
[67] Synge, J. L., *Relativity: The General Theory* (North-Holland, Amsterdam, 1960).
[68] Lense, J. and Thirring, H., Phys. Z. 19 (1918) 156; English translation by Mashhoon, B., Hehl, F. W., and Theiss, D. S. in Gen. Rel. Grav. 16 (1984) 711, DOI: 10.1007/BF00762913.
[69] Nordtvedt, K., Phys. Rev. 169 (1968) 1017, DOI: 10.1103/physrev.169.1017.
[70] Einstein, A., Sitzungsber. Preuss. Akad. Wiss. Berlin (Math. Phys.) 1916 (1916) 688.
[71] Ciufolini, I. and Wheeler, J. A., *Gravitation and Inertia* (Princeton University Press, Princeton, 1995).
[72] Isaacson, R. A., Phys. Rev. 166 (1968) 1263, DOI: 10.1103/PhysRev.166.1263.

[73] Deser, S., Gen. Rel. Grav. 1 (1970) 9, DOI: 10.1007/BF00759198.
[74] Feynman, R. P., Moringo, F. B., and Wagner, W. G., *Feynman Lectures on Gravitation* (Taylor & Francis, Boca Raton, 2003).
[75] Finn, L. S. and Chernoff, D. F., Phys. Rev. D47 (1993) 2198, DOI: 10.1103/PhysRevD.47.2198.
[76] Low, R. J., Class. Quant. Grav. 16 (1999) 543, DOI: 10.1088/0264-9381/16/2/016.
[77] Carlip, S., Phys. Lett. A267 (2000) 81, DOI: 10.1016/S0375-9601(00)00101-8.
[78] Weber, J., Phys. Rev. 117 (1960) 306, DOI: 10.1103/PhysRev.117.306.
[79] Hulse, R. A. and Taylor, J. H., Astrophys. J. 195 (1975) L51, DOI: 10.1086/181708.
[80] Taylor, J. H. and Weisberg, J. M., Astrophys. J. 345 (1989) 434, DOI: 10.1086/167917.
[81] LIGO Scientific and Virgo Collaborations (Abbott, B. P. et al.), Phys. Rev. Lett. 116 (2016) 061102, DOI: 10.1103/PhysRevLett.116.061102.
[82] Bardeen, J. M., Carter, B., and Hawking, S. W., Commun. Math. Phys. 31 (1973) 161, DOI: 10.1007/BF01645742.
[83] Birkhoff, G. D., *Relativity and Modern Physics* (Harvard University Press, Cambridge, 1923).
[84] Oppenheimer, J. R. and Snyder, H., Phys. Rev. 56 (1939) 455, DOI: 10.1103/PhysRev.56.455.
[85] Font, J. A., Living Rev. Relativ. 11 (2008) 7, DOI: 10.12942/lrr-2008-7.
[86] Eddington, A. S., Nature 113 (1924) 192, DOI: 10.1038/113192a0.
[87] Finkelstein, D., Phys. Rev. 110 (1958) 965, DOI: 10.1103/PhysRev.110.965.
[88] Kruskal, M. D., Phys. Rev. 119 (1960) 1743, DOI: 10.1103/PhysRev.119.1743.
[89] Szekeres, G., Publ. Math. Debrecen 7 (1960) 285; reprinted in Gen. Rel. Grav. 34 (2002) 2001, DOI: 10.1023/A:1020744914721.
[90] Carter, B., Phys. Rev 141 (1966) 1242, DOI: 10.1103/PhysRev.141.1242.
[91] Penrose, R., in *Relativity, groups and topology*, edited by DeWitt, B. S. and DeWitt, C. (Gordon and Breach, New York, 1964); reprinted in Gen. Rel. Grav. 43 (2011) 901, DOI:10.1007/s10714-010-1110-5.
[92] Frolov, V. P. and Zelnikov, A., *Black Hole Physics* (Oxford University Press, Oxford, 2011).
[93] Carlip, S., in *One Hundred Years of General Relativity: From Genesis and Empirical Foundations to Gravitational Waves, Cosmology and Quantum Gravity*, edited by Ni, W.-T. (World Scientific, Singapore, 2016), chap. 22.
[94] Gallo, E. and Marolf, D., Am. J. Phys. 77 (2009) 294, DOI: 10.1119/1.3056569.
[95] Narayan, R. and McClintock, J. E., in *General Relativity and Gravitation: A Centennial Perspective*, edited by Ashtekar, A., Berger, B. K., Isenberg, J., and MacCallum, M. (Cambridge University Press, Cambridge, 2015), chap. 3.
[96] Cornish, N. J., Spergel, D. N., and Starkman, G. D., Class. Quant. Grav. 15 (1998) 2657, DOI: 10.1088/0264-9381/15/9/013.
[97] Weinberg, S., *Cosmology* (Oxford University Press, Oxford, 2008).
[98] Dodelson, S., *Modern Cosmology* (Academic Press, San Diego, 2003).
[99] Wolf, J. A., *Spaces of Constant Curvature* (AMS Chelsea Publishing, Providence, 2011).
[100] Luminet, J.-P., Scholarpedia, 10 (2015) 31544, DOI: 10.4249/scholarpedia.31544

[101] Friedmann, A., Z. Phys. 10 (1922) 377.
[102] Lemaître, G., Ann. Soc. Sci. Bruxelles A47 (1927) 49.
[103] Robertson, H. P. Astrophys. J. 82 (1935) 284.
[104] Walker, A. G., Proc. Lond. Math. Soc. s2-42 (1936) 90 DOI: 10.1112/plms/s2-42.1.90.
[105] Longair, M., *The Cosmic Century* (Cambridge University Press, Cambridge, 2006).
[106] Spradlin, M., Strominger, A., and Volovich, A., in *Unity from Duality: Gravity, Gauge Theory and Strings*, edited by Bachas, C. P., Bilal, A., Douglas, M. R., Nekrasov, N. A., and David, F. (Springer, Berlin, 2002), p. 423, DOI: 10.1007/3-540-36245-2.
[107] Gibbons, G. W., in *Mathematical and Quantum Aspects of Relativity and Cosmology*, edited by Cotsakis, S. and Gibbons, G. W. (Springer, Berlin, 2000), p. 102, DOI: 10.1007/3-540-46671-1.
[108] Salisbury, D. C., J. Phys. Conf. Series 222 (2010) 012052, DOI: 10.1088/1742-6596/222/1/01252.
[109] Arnowitt, R. L., Deser, S., and Misner, C. W., in *Gravitation: an introduction to current research*, edited by Witten, L. (Wiley, Hoboken, 1962), chap. 7; reprinted in Gen. Rel. Grav. 40 (2008) 1997, DOI: 10.1007/s10714-008-0661-1.
[110] Dirac, P. A. M., Can. J. Math. 2 (1950) 129, DOI: 10.4153/CJM-1950-012-1.
[111] Henneaux, M. and Teitelboim, C., *Quantization of Gauge Systems* (Princeton University Press, Princeton, 1992).
[112] Hojman, S. A., Kuchar, K., and Teitelboim, C., Annals Phys. 96 (1976) 88, DOI: 10.1016/0003-4916(76)90112-3.
[113] York, J. W., Phys. Rev. Lett. 28 (1972) 1082, DOI: 10.1103/PhysRevLett.28.1082.
[114] Isenberg, J., in *The Springer Handbook of Spacetime*, edited by Ashtekar, A. and Petkov, V. (Springer, Berlin, 2014), p. 303.
[115] Gourgoulhon, E., *3+1 Formalism in General Relativity* (Springer, Berlin, 2012).
[116] Regge, T. and Teitelboim, C., Annals Phys. 88 (1974) 286, DOI: 10.1016/0003-4916(74)90404-7.
[117] Stephani, H., Kramer, D., MacCallum, M., Hoenselaers, C., and Herlt, E., *Exact Solutions to Einstein's Field Equations* (Cambridge University Press, Cambridge, 2003).
[118] Alcubierre, M., *Introduction to 3+1 Numerical Relativity* (Oxford University Press, Oxford, 2008).
[119] Baumgarte, T. W. and Shapiro, S. L., *Numerical Relativity* (Cambridge University Press, Cambridge, 2010).
[120] Hawking, S. W. and Ellis, G. F. R., *The large scale structure of space-time* (Cambridge University Press, Cambridge, 1973).
[121] Geroch, R. and Horowitz, G.T., in *General Relativity: An Einstein Centenary Survey*, edited by Hawking, S. W. and Israel, W. (Cambridge University Press, Cambridge, 1979), chap. 5.
[122] Choquet-Bruhat, Y., *General Relativity and the Einstein Equations* (Oxford University Press, Oxford, 2009).
[123] *General Relativity and Gravitation: A Centennial Perspective*, edited by Ashtekar,

A., Berger, B. K., Isenberg, J., and MacCallum, M. (Cambridge University Press, Cambridge, 2015).
[124] Friedman, J. L., Schleich, K., and Witt, D. M., Phys. Rev. Lett. 71 (1993) 1486, Erratum: Phys. Rev. Lett. 75 (1995) 1872, DOI: 10.1103/PhysRevLett.71.1486, 10.1103/PhysRevLett.75.1872.
[125] Brans, C. and Dicke, R. H., Phys. Rev. 124 (1961) 925, DOI: 10.1103/PhysRev.124.925.
[126] Horndeski, G. W., Int. J. Theor .Phys. 10 (1974) 363, DOI: 10.1007/BF01807638.
[127] Gleyzes, J., Langlois, D., Piazza, F., and Vernizzi, F., Phys. Rev. Lett. 114 (2015) 211101, DOI: 10.1103/PhysRevLett.114.211101.
[128] de Rham, C., Living Rev. Relativ. 17 (2014) 7, DOI: 10.12942/lrr-2014-7.
[129] Berti, E. et al., Class. Quant. Grav. 32 (2015) 243001, DOI: 10.1088/0264-9381/32/24/243001.
[130] DeWitt, B. S., Phys. Rev. 160 (1967) 1113, DOI: 10.1103/PhysRev.160.1113.
[131] DeWitt, B. S., Phys. Rev. 162 (1967) 1195, DOI: 10.1103/PhysRev.162.1195.
[132] DeWitt, B. S., Phys. Rev. 162 (1967) 1239, DOI: 10.1103/PhysRev.162.1239.
[133] Kiefer, C., *Quantum Gravity* (Oxford University Press, Oxford, 2012).
[134] Carlip, S., Rep. Prog. Phys. 64 (2001) 885, DOI: 10.1088/0034-4885/64/8/301.
[135] Polchinski, J., *String Theory* (Cambridge University Press, Cambridge, 2005).
[136] Elvang, H. and Horowitz, G. T., in *General Relativity and Gravitation: A Centennial Perspective*, edited by Ashtekar, A., Berger, B. K., Isenberg, J., and MacCallum, M. (Cambridge University Press, Cambridge, 2015), chap. 12.
[137] Rovelli, C., *Quantum Gravity* (Cambridge University Press, Cambridge, 2004).
[138] *Approaches to Quantum Gravity*, edited by Oriti, D. (Cambridge University Press, Cambridge, 2009).
[139] Sakstein, J., "Astrophysical Tests of Modified Gravity," Ph.D. thesis, University of Cambridge, June 2014, eprint arXiv:1502.04503.
[140] Turyshev, S. G., Nucl. Phys. Proc. Suppl. 243-244 (2013) 197, DOI: 10.1016/j.nuclphysbps.2013.09.024.
[141] Thurston, W. P., *Three-Dimensional Geometry and Topology* (Princeton University Press, Princeton, 1997).
[142] Morgan, J. W., Bull. Amer. Math. Soc. 42 (2005), 57, DOI: 10.1090/S0273-0979-04-01045-6.
[143] Brans, C. H., Astrophys. Space Sci. Libr. 211 (1997) 160, DOI: 10.1007/978-94-011-5812-1_27.

Index

acceleration 1–2
  of expansion of the Universe 95–96
  relative 3, 47, 52
action
  boundary terms in 107–108
  Brans-Dicke 112
  effective 114
  Einstein-Hilbert *see* Einstein-Hilbert action
  electromagnetic field 62, 106
  Hamiltonian form 104, 108
  Lovelock 58
  matter 54, 56–57
  Palatini 58
  point particle 60
  reduced phase space 128
  scalar field 61–62
  symmetries of 56–57
action principle 52, 53
ADM metric 101–102
AdS *see* anti-de Sitter space
AdS/CFT correspondence 111, 114
affine parameter 13, 126
alternating symbol *see* Levi-Civita symbol
alternative models 23–24, 57–58, 112, 116
anti-de Sitter space 55, 96, 100, 111, 114
Arnowitt, R. 108
asymptotic safety 114
asymptotic structure *see* spacetime, global structure

Bardeen, J. M. 89
Barish, B. C. 79
basis 27–31
  dual 29–31
  orthonormal *see* orthonormal basis
  tensor *see* tensor, basis
  vector *see* vector, basis
basis one-form *see* frame, one-form
Betti number 121
Bianchi identity 48–49, 93
bimetric theories 112
Birkhoff's theorem 81, 91
Birkhoff, G. D. 81
black hole 80–91
  astrophysical 78, 87, 91, 115, 116
  binaries 78, 91, 110, 115
  boundary terms 108
  charged *see* black hole, Reissner-Nordstrom

  dynamical 91
  extreme mass ratio 77, 109
  horizon 84–91
    as event horizon 86, 87, 89
    as Killing horizon 90–91
    as marginally trapped surface 90
  Kerr 89, 109
  Kruskal-Szekeres extension 86–87
  observations 74, 77, 78, 115, 116
  Penrose diagram 88
  quasinormal modes 70, 77, 78
  Reissner-Nordstrom 89, 111
  rotating *see* black hole, Kerr
  surface gravity 86, 89
black hole mechanics 89, 91
black hole thermodynamics 89, 91, 114, 116
BMS *see* Bondi-Metzner-Sachs group
Bondi-Metzner-Sachs group 112
boundary *see* manifold, with boundary
Brans-Dicke theory 112, 116

Cartan calculus 38–39, 50–51, 121–123
  for Schwarzschild metric 82–84
  integration 122–123
Cartan structure equations 50, 82–83
Cartan, E. 39
Carter, B. 89
Carter-Penrose diagram *see* Penrose diagram
Cartesian coordinates 4–5, 7, 27, 28, 45, 46
  asymptotic 107
Cassini spacecraft 24
Cauchy horizon 111
causal dynamical triangulations 114
causal sets 114
causal structure 8–9, 88, 99
Choquet-Bruhat, Y. 110
Christoffel connection *see* connection, Levi-Civita
Christoffel symbols *see* connection, Levi-Civita
Christoffel, E. B. 42
CMBR *see* cosmology, CMB
CMC *see* time-slicing, constant mean curvature
cohomology
  de Rham 39, 121–122, 128
conformal transformation *see* Weyl transformation
connection 39–44
  as independent variable 57–58

connection (*continued*)
  basis change 41
  Levi-Civita 13, 35, 42–44, 47, 53
    linearized 66
  metric compatible 42, 43
  one-form 50
    for Schwarzschild metric 82
    in two dimensions 51
  spin 50
conservation laws 56–57, 62–65
  from invariance 56–57, 59
  in cosmology 95
  integral 62–63
  and Killing vectors 63
constraints 104–105, 108, 110
  and gauge invariance 105–106, 108
  diffeomorphism 104
  Dirac analysis 105, 108
  electromagnetism 106
  first class 105
  Hamiltonian 104, 110
    boundary terms 107–108
    quantum 113
  integrable 107, 108
  momentum 104, 110
  smeared 105–108
continuity equation 62, 64
contravariant 30
conventions 14
  curvature tensor 49, 55
  derivatives and covariant derivatives 40
  indices 29, 33–34
  metric signature 10, 55
  quadrupole moment 74
  sign of action 55
  sign of stress-energy tensor 55
  summation 10
  three- and four-dimensional notation 103
coordinate independence 7, 16, 22, 24, 26–27, 37, 82, *see also* diffeomorphism invariance
coordinate patch *see* manifold, coordinate patch
coordinate transformation 28, 31, 33, 37, 122
  active and passive 26, 119
coordinates 26
  adapted to an isometry 63, 80, 126
  areal 82
  Cartesian *see* Cartesian coordinates
  comoving 93–94
  conformal time 93, 99
  cosmological time 93–94
  Eddington-Finkelstein 85–86, 91
  Gaussian normal 82
  isotropic 16, 23, 82
  Kruskal-Szekeres 86–87, 90, 91
  periodic 121
  polar *see* polar coordinates
  proper distance 82
  proper time 82, 93
  radial 82
  Riemann normal 54
  spherical *see* spherical coordinates
  synchronous 82
  tortoise 85
  York time *see* York decomposition
cosmic censorship 110–111
  strong 111
  weak 110
cosmic microwave background *see* cosmology, CMB
cosmological constant 54, 61, 95, 96
cosmology 92–100
  CMB 94, 97–116
    fluctuations 98–99
    last scattering 97, 99
  CMB 98
    fluctuations 98
  comoving coordinates 93–94
  conservation 95
  density parameter 96
  expansion of the Universe 92, 94, 95, 97
    acceleration 95–96
  FLRW metric 94–96, 100
    Penrose diagram 99
    predictions 96–98
  Friedmann equations 95
  homogeneity 92–93
  horizon 98–99
  Hubble constant 95
  inflation 58, 98–100
    inflaton 100
  initial singularity 95, 99, 109, 110
  isotropy 92–93
  primordial nucleosynthesis 97
  red shift 97
  scale factor 94
  structure formation 98, 116
  thermal history 97–98
  topology *see* topology, of the Universe
cotangent space 30
covariant 30
covariant derivative *see* derivatives, covariant
curvature
  extrinsic 102–103, 106, 127
  intrinsic 25, 102–103
  mean *see* mean curvature
curvature scalar 48, 50, 53
  3+1 decomposition 103
  spatial 103, 107, 127
curvature tensor 45–49
  and covariant derivatives 46
  and geodesic deviation 46–47
  and parallel transport 46, 123–126
  contractions 48–49
  conventions 49
  defined 45
  extrinsic 102–103
  spatial 93
  symmetries 47–48
  two-form 50

for Schwarzschild metric 82–83
in two dimensions 51
curve
  autoparallel 4, 12, 42
  geodesic *see* geodesic
  holonomy of 46, 123–126
  lightlike *see* curve, null
  noncontractible 126
  null 8–9, 88, *see also* geodesic, null
  parametrized 4, 27–28, 44, 46, 118, 123–125
  path ordering 124
  spacelike 8–9
  timelike 8–9

dark energy 95, 96, 112, 116
dark matter 96, 112, 116
de Donder gauge *see* weak field approximation, de Donder gauge
de Sitter precession *see* geodetic precession
de Sitter space 55, 96, 100
density (in mathematics)
  basis change 38, 122
  scalar 37–38, 44, 122
  tensor 122
density (in physics) *see* energy, density
derivatives
  covariant 35, 39–41
    commutator 42, 46
    divergence of a tangent vector 43
  exterior 38–39
  Lie 126–128
  product rule 13, 39–41
Deser, S. 108
DeWitt metric 113
DeWitt, B. S. 116
diffeomorphism 25, 118–119
  and Lie derivative 127
  and surface deformations 105
  defined 118
diffeomorphism invariance 26–27, 33, 56–57, 65, 119, 121, *see also* coordinate independence
  and background structures 26, 36, 53, 113
  and observables 113
differential form 30, 36, 50–51, 121–123, *see also* Cartan calculus
  closed 121–122
  defined 36
  exact 121–122
  exterior derivative *see* derivatives, exterior
  harmonic 123
  Hodge decomposition 123
  Hodge dual 36, 122–123
  integration 122–123
  Laplacian 123
  wedge product 36
differential structure 121, 128
Dirac, P. A. M. 105, 108
dual space 29–30

East Coast *see* conventions, metric signature
Eddington, A. S. 20, 24
Eddington-Robertson-Schiff parameters 24
Einstein field equations 52–56, 58
  approximations 13–14, 64, 109–110, *see also* weak field approximation
  coupling constant 54, 56, 68
  degrees of freedom 75–76, 105–107, 127–128
  exact solutions 109, 116, *see also* Schwarzschild metric; cosmology, FLRW metric
  from action *see* Einstein-Hilbert action
  from flat spacetime 71
  Hamiltonian form 104
  initial value problem 75, 106–107, 110, 116, 127–128
  Newtonian limit 13–14, 52, 68
  numerical solution *see* numerical relativity
  observational tests *see* observations
  trace reversed 56
  weak fields *see* weak field approximation
Einstein tensor 48–49, 56
  in cosmology 94–95
  linearized 67
Einstein, A. 1, 8, 15, 23, 53, 58, 67, 71, 112
Einstein-Hilbert action 53–56, 58
  from flat spacetime 71
  generalizations 57–58, 112
electromagnetic field 62, 106
energy
  density 59, 61, 62, 95, 97
  gravitational 63, 65, 69–71
    pseudotensor 63, 65
    quasilocal 63, 65, 108
energy conditions 110
energy tensor *see* stress-energy tensor
equation of state 61, 83, 95–96
equations of motion, *see also* geodesic equation
  from field equations 63–65
equivalence principle *see* principle of equivalence
event horizon 86, 87, 89
exact solutions *see* Einstein field equations, exact solutions
exterior derivative *see* derivatives, exterior
extrinsic curvature *see* curvature, extrinsic
Eötvös, L. 3

fiber bundle 51
FLRW *see* cosmology, FLRW metric
Fock gauge *see* weak field approximation, de Donder gauge
frame 29
  cosmological 92
  freely falling 3, 54, 55, 63
  one-form 50
    for Schwarzschild metric 82
    in two dimensions 51

frame (*continued*)
  orthonormal 35–36, 45
  preferred 24, 53
  rest 61, 76, 92
frame dragging 69, 116
Friedmann, A. 100
Frobenius's theorem 128
FRW *see* cosmology, FLRW metric
fundamental group 126

Galileo 2
Gauss, C. F. 103
Gauss-Codazzi equation 103
general covariance 27, 119, *see also* diffeomorphism invariance
geodesic deviation 46–47, 52, 76–77
  Newtonian 52
geodesic equation 12–14, 42
  Einstein-Infeld-Hoffmann form 64, 65
  first integral 13
  from field equations 63–65
  from stress in singular strut 64, 65
geodesics 4–14
  defined 4
  determine geometry 43
  great circle 6
  in cosmology 93
  Newtonian limit 13–14
  null 43, 49, 75, 84–86, 88, 90
  Schwarzschild 15–24
geodetic precession 116
geometrized units 56
geometry 2–3, 25
  constant curvature 93, 100, 120
  differential 25–51, 117–128
  Euclidean 4
  from geodesics 4, 43
  global 100, 110, 116
  pseudo-Riemannian 10, 42, 53
  Riemannian 10, 42
GPS 23, 24
gradient *see* vector, cotangent, gradient
gravitational collapse 79, 83, 87, 91, 110
gravitational energy *see* energy, gravitational
gravitational lensing *see* light, deflection
gravitational mass 1–3, 59, 70
gravitational radiation *see* gravitational waves
gravitational red shift 21–23, 97
gravitational time dilation 23–24, 115
gravitational waves 72–79
  chirp 78
  detectors 76–77, 79
    bar detectors 77
    interferometers 77
  observation *see* observations, gravitational waves
  polarization 75, 106, 115
  power 73–74, 108
  primordial 79, 116
  propagation 74–76

quadrupole radiation 72–73, 79
radiation reaction 64, 70, 74, 79, 116
sources 77–79
speed of 75, 79, 112, 115
weak field solution 72–76
  radiation gauge 74–76, 79
gravitomagnetism 69, 71
  observation *see* observations, frame dragging
Gravity Probe A 23
Gravity Probe B 69

Hamiltonian 84, 104, 105
  ADM 107–108
  electromagnetism 106
  quantum 113
Hamiltonian formalism 101–108
  action 104
  boundary terms 107–108
  canonical momentum 104
  constraints *see* constraints
  electromagnetism 106
  Poisson brackets 104
  raising and lowering indices 102, 103
harmonic gauge *see* weak field approximation, de Donder gauge
Hawking radiation 89, 91
Hawking, S. W. 89, 95, 110
Hilbert, D. 53, 58
Hodge decomposition 123
Hodge dual 36, 122–123
holonomy 46, 123–126
homeomorphism 25, 117, 119, 121
  defined 25
horizon *see* black hole, horizon; cosmology, horizon; Cauchy horizon
Horndeski theories 112, 116
Hulse, R. A. 79
hypersurface 9, 126
  constant time 92–94, 101–103, 105
  null 9, 89–91
  spacelike 9, 92–94
  timelike 9

Icarus 19
impact parameter 19, 21
indices 10–11, 32, 33
  conventions 29, 33–34, 103
  dummy 11
  permutations 32
  raising and lowering 34–35, 46, 66, 70, 102
inertial mass 1–3, 70
inflation 58, 98–100
initial value problem 75, 106–107, 110, 116, 127–128
integrals 37–38, 122–123
  boundary terms 44, 55, 62–63, 107–108, 123
  integration by parts 43–44, 107–108, 123
  Jacobian 37, 122

Stokes' theorem 39, 43–44, 122–123, 128
interior solution *see* metric, interior
isometry 34–35, 80–82, 126–128
   adapted coordinates 63, 80, 126
   asymptotic 111–112
   defined 34

Jacobi identity 48
Jacobi's formula 42, 55
Jacobian *see* integrals, Jacobian

Killing equation 34–35, 63, 80, 91, 126
Killing horizon 91
Killing vector 34, 63, 80–81, 90–91, 126–128
Killing, W. 34
Kretschmann scalar 84
Kretschmann, E. 26, 36
Kronecker delta *see* tensor, Kronecker delta

LAGEOS 19, 69
Laplace, P.-S. 75
lapse function 101, 105
Leibniz rule *see* derivatives, product rule
Lemaître, G. 100
Lense, J. 69
Lense-Thirring effect 69
Levi-Civita connection *see* connection, Levi-Civita
Levi-Civita symbol 32, 93, 122
Levi-Civita tensor 122–123
Levi-Civita, T. 42
Lichnerowicz operator 70
Lichnerowicz-York equation 107, 127–128
Lie derivative *see* derivatives, Lie
light
   deflection 19–20, 24, 116
   time delay 20–21, 24, 116
light cone 9, 43, 84–86, 88, 99
lightlike *see* curve, null; hypersurface, null
LIGO 77, 79, 91
line element 4, 6, 8, 10, 12–14, 35, *see also* metric
local Lorentz invariance 3, 24, 36
loop quantum gravity 114, 116
Lorentz, H. 67
Lorenz gauge *see* weak field approximation, de Donder gauge
Lorenz, L. 67
Lovelock action 58
Lunar laser ranging 24, 70, 116

manifold 25–27, 117–118
   coordinate patch 25–26, 37, 117–118
   defined 25–26, 117–118
   extension 86
   maps between 118–119
   pull-back 119
   push-forward 118–119
   submanifold 9, 102, 103, *see also* hypersurface

   topology *see* topology
   transition function 25, 117–118
   with boundary 26, 107, 118
marginally trapped surface *see* horizon, as marginally trapped surface
Mars 19, 116
mass
   ADM 108
   Bondi 108
   gravitational 1–3, 59, 70
   in Schwarzschild metric 15, 84, 108
   inertial 1–3, 70
massive gravity 112, 116
mathematical relativity 110–112, 116
mean curvature 103, 106, 127
Mercury 18–19, 21, 24, 116
metric 9–11, 33–35
   ADM *see* ADM metric
   and angles 44
   and lengths 44
   anti-de Sitter *see* anti-de Sitter space
   asymptotically flat 15, 107, 108, 112
   constant curvature *see* geometry, constant curvature
   de Sitter *see* de Sitter space
   defined 10, 33
   determinant 34, 37–38
   flat 7, 8, 45, 46, 66, 71, *see also* Minkowski metric
   FLRW 94–96, 100
   in orthonormal basis 35–36, 50
   in Riemann normal coordinates 54
   interior 83, 87, 91, 109
   inverse 11, 33
   Lorentzian 11, 38
   Minkowski *see* Minkowski metric
   Newtonian limit 13–14, 68
   perturbation 66
      trace reversed 67
   PPN 23
   raising and lowering indices with 34–35, 46
   Riemannian 11, 34
   Schwarzschild *see* Schwarzschild metric
   signature 10, 11
   spherically symmetric 81–82, 126, 128
   static 80–81, 126
   stationary 80–81, 126
   symmetries *see* isometries
   Yamabe 107, 127
minimal coupling 61
Minkowski metric 8, 15, 61, 66, 87–88
Minkowski, H. 8
Misner, C. W. 108

natural units 56
neutron stars 78, 110, 115, 116
   binaries 74, 78
Newton, I. 1
Newtonian gravity 1–3, 13–14, 52, 68
   multipole expansion 72

Newtonian gravity (*continued*)
  potential 14, 21, 52, 68
  speed of 75
  trajectories 18
Noether's theorem 57, 58, 62
Noether, E. 57
nonmetricity 42, 57, 112
  and geodesics 42
Nordtvedt effect 70, 71, 116
null *see* curve, null; hypersurface, null
numerical relativity 77, 109–110, 116

observations 24, 116
  alternative models 112, 116
  black holes 78, 91, 115, 116
  cosmology 96–98, 116
  deflection of light 20, 24, 116
  frame dragging 69, 71, 116
  geodetic precession 116
  gravitational red shift 23–24
  gravitational waves 74, 77, 78, 115–116
  inverse square law 115
  Nordtvedt effect 70, 71, 116
  perihelion precession 18–19, 24, 116
  principle of equivalence 2, 3, 115–116
  Shapiro time delay 21, 24, 116
  time dilation 23–24, 115
Oppenheimer, J. R. 91
orbits 18–19, 70, 75, 110, 116
  decay 74, 78, 79
  precession 18–19, 24
orthonormal basis 35–36, 50–51
Ostrogradsky's theorem 2–3

Palatini variational principle *see* action, Palatini
parallel transport 44, 124–125
  and covariantly constant vectors 45
  and curvature 46, 123–126
  preserves lengths and angles 44
parallel transport matrix 125
parametrized post-Newtonian *see* PPN
path ordering operation 124
Penrose diagram 49, 87–88, 91, 99, 111
Penrose, R. 95, 110
Petrov classification 109
Planck force 77
Planck length 113
polar coordinates 5–6, 28, 84
Pound, R. 23
Poynting vector 62
PPN formalism 23–24, 109
Pretorius, F. 109
principle of equivalence 1–3, 43, 65, 70
  Einstein 3
  observational tests 2, 3, 115–116
  strong 3
  weak 3
projective structure 43
pull-back 119
pulsars 19, 24, 74, 79, 115, 116

Hulse-Taylor 74, 79, 116
push-forward 118–119

quadrupole moment 72–74, 78
quantum gravity 89, 99, 112–114, 116
  loop quantum gravity 114, 116
  observables 113
  problem of time 113
  string theory 114, 116
  tests 115, 116
quasinormal modes *see* black hole, quasinormal modes

radiation reaction *see* gravitational waves, radiation reaction
Raychaudhuri equation 48
Rebka, G. 23
reduced phase space 128
Ricci tensor 48, 54
  for Schwarzschild metric 83–84
  linearized 66
  variation 55
Ricci-Curbasto, G. 48
Riemann curvature *see* curvature tensor
Riemann normal coordinates 54
Riemann, B. 45
Robertson, H. P. 100

scalar curvature *see* curvature scalar
scalar density *see* density, scalar
scalar field 61–62, 100, 112
scalar-tensor theories *see* alternative models
Schwarzschild metric 15–24, 35, 82–91
  and black holes *see* black holes
  geodesics 15–24
Schwarzschild, K. 15
Shapiro time delay 20–21, 24, 116
Shapiro, I. 21
shift vector 101, 105
singularities 84, 110–111
  "Schwarzschild singularity" (horizon) 84
  and cosmic censorship 110–111
  coordinate 84
  curvature 84, 86
  future 87
  indicated in spacetime diagram 87
  initial 95, 99, 109
  past 87
  theorems 95, 100, 110, 116
  timelike 111
Snyder, H. 91
spacelike *see* curve, spacelike; hypersurface, spacelike
spacetime 3, 8–10
  global structure 87–88, 100, 110–112, 116
special relativity 3, 8–9, 14, 59, 71
speed of gravity 75, 79, 112, 115
sphere *see* topology, examples
spherical coordinates 6, 15, 81
spin connection *see* connection, spin

spin two 71, 76
spinor 50, 58
Starobinsky, A. 58
Stokes' theorem *see* integrals, Stokes' theorem
stress-energy tensor 54, 59–65, *see also* energy, gravitational
  and equations of motion 63–65
  and gravitomagnetism 69
  and Noether's theorem 65
  as a rank two tensor 59
  as a variational derivative 54
  conservation *see* conservation laws
  dust 61
  electromagnetic field 62
  Newtonian limit 68
  perfect fluid 61, 65, 83, 94
  point particle 60
  scalar field 61–62, 100
  sign convention 55
string theory 114, 116
supernovae 79, 110
surface deformations 105, 108
surface gravity 86, 89

tangent space 28
Taylor, J. H. 79
tensor 31–33
  antisymmetrization 32
  basis 32–33
  basis change 33, 37, 122
  components 31
  contraction 33
  curvature *see* curvature tensor
  defined 31
  Einstein *see* Einstein tensor
  Kronecker delta 31
  Levi-Civita *see* Levi-Civita tensor
  metric *see* metric
  nonmetricity *see* nonmetricity
  rank 31
  Ricci *see* Ricci tensor
  stress-energy *see* stress-energy tensor
  symmetrization 32
  torsion *see* torsion
  type 31
  Weyl *see* Weyl tensor
tensor field 31
tensor product 32
tetrad 29
theorema egregium 103
Thirring, H. 69
Thorne, K. S. 79
time dilation *see* gravitational time dilation
time-slicing 101
  constant mean curvature 106, 110, 127
timelike *see* curve, timelike; hypersurface, timelike

topology 36, 119–122, 128
  Betti number 121
  cohomology 39, 121–122, 128
  examples 119–121
  fundamental group 126
  of the Universe 94, 100
  of three-manifolds 36, 94, 121, 128
torsion 42, 57, 58, 112
torus *see* topology, examples

universal coupling *see* principle of equivalence

vector
  as a derivative 27–28, 118
  basis 28
    anholonomic 28
    coordinate 28, 30
    dual *see* basis, dual
    holonomic 28
    orthonormal *see* orthonormal basis
  basis change 28, 31
  cotangent 29–31, 119
    gradient 30
  covariantly constant 45–46
  Killing *see* Killing vector
  pull-back 119
  push-forward 118–119
  tangent 27–28, 31, 118
vector field 28
Venus 19, 21
vielbein 29
vierbein 29
volume element 38, 48, 54

Walker, A. G. 100
weak field approximation 66–71, 116
  curved backgrounds 70–71
  de Donder gauge 67
  gauge choice 67, 71
  higher orders 69, 71, 109
  radiation gauge 74–75, 79
  raising and lowering indices 66, 70
Weber, J. 77
wedge product *see* differential form, wedge product
Weiss, R. 79
West Coast *see* conventions, metric signature
Weyl tensor 49
Weyl transformation 43, 49, 88
Weyl, H. 49
Wheeler, J. A. 25
Wheeler-DeWitt equation 113, 116
white dwarfs 23, 116
white hole 87
world line 9, 61

York decomposition 106–108, 127–128

The manufacturer's authorised representative in the EU for product safety is
Oxford University Press España S.A. of el Parque Empresarial San Fernando de
Henares, Avenida de Castilla, 2 – 28830 Madrid (www.oup.es/en or product.
safety@oup.com). OUP España S.A. also acts as importer into Spain of products
made by the manufacturer.

www.ingramcontent.com/pod-product-compliance
Ingram Content Group UK Ltd.
Pitfield, Milton Keynes, MK11 3LW, UK
UKHW051651180426
11946UKWH00005B/111